RENEWALS 458-4574

Understanding Technological Innovation

Understanding Technological Innovation

A Socio-Technical Approach

Patrice Flichy

*Professor, Department of Sociology (LATTS),
University of Paris-Est, France*

Translated by Liz Carey-Libbrecht

English translation of *L'Innovation technique*, Paris, La Découverte

Edward Elgar
Cheltenham, UK • Northampton, MA, USA

© Éditions La Découverte, Paris, 1995, 2003
© Patrice Flichy for the English Version 2007

All rights reserved. No part of this publication may be reproduced, stored in a retrieval system or transmitted in any form or by any means, either electronic, mechanical or photocopying, recording, or otherwise without the prior permission of the publisher.

Published by
Edward Elgar Publishing Limited
Glensanda House
Montpellier Parade
Cheltenham
Glos GL50 1UA
UK

Edward Elgar Publishing, Inc.
William Pratt House
9 Dewey Court
Northampton
Massachusetts 01060
USA

A catalogue record for this book is available from the British Library

Library of Congress Cataloguing in Publication Data
Flichy, Patrice.
 [Innovation technique. English]
 Understanding technological innovation : a socio-technical approach / by Patrice Flichy ; translated by Liz Carey-Libbrecht.
   p. cm.
 "English translation of L'innovation technique."
 Includes bibliographical references and index.
 1. Social sciences—Methodology. 2. Technological innovations. I. Title.
 H61.F54513 2007
 303.48'3—dc22
                                                              2007016042

ISBN 978 1 84720 391 5

Printed and bound in Great Britain by MPG Books Ltd, Bodmin, Cornwall

Contents

Preface *vi*

PART I STANDARD THEORIES

1. From no technology to all technology 3
2. Technological determinism – social determinism 19

PART II A SOCIO-TECHNICAL APPROACH

3. The anthropology of technology: thinking the technological and the social together 43
4. Socio-technical action and frame of reference 73

PART III SOCIO-TECHNICAL HISTORY

5. The time of technology 99
6. Technological *imaginaire* 120
7. The birth of socio-technical frames 145

Conclusion 163

Bibliography *167*
Index *189*

Preface

The motto of the Chicago World Fair in 1933: 'Science finds, industry applies, man conforms', sounds like something from another age. The linear science–technology–use schema no longer works today. Engineers are questioning user-centred design as the benefits of technological progress no longer appear obvious to all and the harmful consequences of technology, or at least the risks it entails, are a prominent topic on the public agenda. Technology assessment procedures urge us to reflect on technological choices. But experts and politicians do not necessarily have the answers that enable us to make wise choices. Other actors, especially ordinary individuals, need to be involved; hence, the recent development of public participation and citizen panels.[1]

Three-quarters of a century after the Chicago World Fair, technology is still one of the kingpins of our societies, but the number of actors who wish to be involved in technological choices is now far greater. In this new context, the social science output on the subject has increased substantially. Social scientists are regularly invited to work with technology assessment offices as well as with industrial research laboratories.

As the social science production in this sphere has grown, it has simultaneously undergone profound change. In standard theories, technology was a black box that the social sciences rarely opened. They studied general frameworks (research and development expenditures), the diffusion of innovations and their effects on daily life or business environments, but showed very little interest in the shaping of technology or even in the details of its uses. When studying technology, social scientists set off from their home ground. They consequently projected their own perspectives onto it: economists of growth examined the role of technology in their model; sociologists of communication saw the diffusion of innovations as an interesting case for analysing the influence of the media; historians created a subdiscipline, the technical history of technology which explored the evolution of technologies over time, without regard for other factors of social life.

From the 1980s, a number of social scientists decided to pay far more attention to technology, to study its different dimensions and to do so in places in which they had rarely undertaken research: the research laboratory, but also the home or the office, where technologies were used on a daily basis. They started to study not only successful objects but also abortive attempts and failures, both objects that were used and those that spent their life in a cupboard. Like physicists who, in exploring matter, try to find increasingly small elementary particles, social scientists started to examine the technical object from the inside, in more and more detail. Behind their approach lay the questions: which choice was made when the technical object was designed, and what do users do with it? The idea was not to draw up the technological genealogy of the one best solution, or to examine how engineers responded to the needs of humans or markets, but rather to examine the processes of trial and error, and the different alternative routes. Finally, some contemporary studies consider that the technical object is not only a functional entity; it also conveys meaning. The object is not only material; it is also symbolic.

All these approaches have marked a profound departure from standard theories. However, most of them still separate the design and production of technologies from their use. To be sure, contemporary research on design does include users; not only is it user-centred, but it also considers that certain users, the lead users, play a crucial part in design (von Hippel, 2002). Many social scientists would therefore agree with a former director of the Xerox research lab that 'design and use mutually shape one another in iterative, social processes' (Brown and Duguid, 1994: 29). As for sociologists who observe uses, whereas they no longer pay attention to the technology, some do observe the relationship with the object in detail: representation, manipulation, choice of functionalities and so on. The question is not whether people use a machine but what they actually do with it. Turkle (1982), for example, asked this question in respect of computers, while Jouët (1994) analysed how the everyday life communication process became increasingly technical. Yet, despite the unquestionable advances of all these research studies, they are still situated either in the sphere of design or in that of use. The aim of this book is to go one step further and to combine the two perspectives within the same analysis.

To present this new approach, it is necessary to discuss the different social science theories on technology. I shall therefore present them in some detail. Of course an entire overview of the social science literature on the subject is not possible within the scope of this book. My intention is critically to examine a range of works with a view to formulating a new approach to the analysis of technological innovation. The choice of authors with whom I establish a dialogue is partly subjective; the order in which I present their work is that which seemed most appropriate for the construction of my approach.

I wish also to point out that the authors in question belong to various disciplines: sociology, history, economics and anthropology. This choice naturally reflects a wish to be situated within the field of social science as a whole. The drawback of such an approach is that it involves the real risk of concepts being borrowed carelessly and taken out of the theoretical framework in which they are meaningful. I nevertheless felt that this risk could be taken because many of the works on which I drew already had a pluridisciplinary dimension associating historians, economists, sociologists and anthropologists. Moreover, my main aim was actually less to situate myself in the all-embracing sphere of social science, than to base the construction of my thesis on boundary concepts that lie at the interface between several disciplines, the fertility of which is a direct result of their heterogeneity.

This reflection on innovation is therefore of a general nature, but the cases used to illustrate my arguments are drawn primarily from the field of information and communication technology (ICT). I have studied old technologies of the nineteenth and early twentieth centuries (Flichy, 1995) as well as new ones, especially videotext (Jouët et al., 1991) and the Internet (Flichy, 2007). I shall use these examples to illustrate my theoretical reflection on innovation, which matured during my studies of these subjects and was fed, to a large extent, by that research.

The first version of this book was published in French. This English version has been amended substantially to give more consideration to the English-language literature on the subject and to take into account research on technological innovation or ICTs published recently[2]. While this volume therefore has a strong international dimension, it nevertheless remains rooted in a French research tradition. This is more specifically the case when I diverge from my own scientific background (the sociology of technology and of ICTs). Hence, the passage that examines how the sociology of work has addressed the subject of technology draws essentially on the French tradition, which is particularly rich in this respect, while the US and British traditions are simply mentioned in passing. Likewise, the rich tradition of feminist, green and development studies has not been included.

The translation from French to English sometimes involves problems of terminology. In French, we clearly distinguish technique from technology. As Leroi-Gourhan (1964) put it, a technique is a combination of gesture and tool. It always associates the object or machine with the human being who builds and uses it. Technology, by contrast, is the study and knowledge of techniques. In English the word 'technology' tends to embrace both meanings.

The route followed in this book consists of three steps. In Part I, I present standard theories on technology, organized around a clear division between technology and society. While some studies focus on laboratories and inventors, others concentrate on the diffusion of a technology and its uses, without the two ever meeting. More significantly, while some scholars examine the conditions of innovation and others its diffusion, the technology itself remains excluded; it is a black box into which nobody dares to look. When, in other theories, a link is established between technology and society, this relationship is always seen from a deterministic viewpoint. The main question is then to know which of the two determines the other or, at least, influences it.

Alongside these largely dominant theories in economics, sociology and history, new approaches have been proposed by anthropology and the new sociology of science and technology (Social Construction of Technology: SCOT; and Actor Network Theory: ANT). They try to open the black box and to consider both technological and social questions. This is the theme of Part II. After presenting this research tradition, with studies of use of ICTs in everyday life and in the workplace, and drawing upon the contribution of interactionist sociology, I present my own approach to innovation. I study how the socio-technical action of the different actors in innovation, particularly designers and users, is organized within the same frames of reference. The most original aspect of this position is that it presents an approach which focuses as much on the design of technologies as on their use.

Part III introduces a new element into the model: time. What part does it play and how is it involved in technological choices? This analysis concerns not only the series of technological events studied in an evolutionary approach or in a switch of paradigm, and the irreversibility of choices made, but also the evolution of representations and of technological and social utopias. Social imagination or *imaginaire* plays an important role in the shaping of innovation. With these basic elements we are able to study the different phases through which both the functioning and the uses of an innovation are developed.

NOTES

1. See the special issue of *Science and Public Policy*, 1999, **26** (5).
2. For bibliographic references and citations, the English-language versions of English-language publications are always used. In the case of French publications, we have translated the citations ourselves. If an English version of the publication exists, it is noted in the bibliography.

PART I

Standard theories

1. From no technology to all technology

In one of his books, Bruno Latour explains the failure of certain technical projects by the lack of their promoters' love for their inventions (Latour, 1992b). But if engineers can be said not to love technology enough, intellectuals can often be considered not to like it at all and sociologists to overlook it entirely. Admittedly, recent decades have witnessed the birth of journals like *Technology and Culture* (1959) and *Culture technique* (1979), but on the whole the social sciences leave little room for technology. Economists usually exclude it from their field of interest: after sifting through economic facts with their concepts (production, capital, labour and so on), they call the remainder 'technology'. Sociologists often consider technology to be a black box: they observe its diffusion and focus mainly on its effects. As for historians, those who show an interest in the subject tend to enclose it in a private territory, so that the technical history of technology completely disregards society. Other intellectuals see technology as relating to modernity; it encompasses society as a whole, not through its objects but through its principles of organization of work and consumption. In the final analysis, the traditional position of the social sciences is either to ignore technology or to see it as a foregone conclusion. I consider that both these attitudes, examined in this chapter, are grounded in the same mistrust.

THE RESIDUAL FACTORS IN ECONOMIC ANALYSIS

Economists have traditionally been interested in technological progress and its impact on employment. Adam Smith saw mechanization and the division of labour as generators of wealth. David Ricardo and then Karl Marx, on the other hand, believed that by substituting capital for labour, new machines caused unemployment. Neo-classical economists turned away from economic development to study the general equilibrium in an essentially static context. In this new theoretical framework the question of technology was largely overlooked, so that technology was in effect excluded from the scope of economics. Lionel Robbins, a prominent London academic, wrote before the Second World War: 'The problem of technique and the problem of economy are fundamentally different problems ... Economists are not interested in technique as such. They are interested in it solely as one of the influences determining relative scarcity' (Robbins, 1932: 35 and 37).

Austrian economist Friedrich von Hayek believed:

> [A]s long as it concerns his problem, the engineer does not participate in a social process in which others can take independent decisions; he lives in his own separate world. ... He does not have to seek available resources nor know the relative importance of various needs. ... He has been given knowledge on the properties of things that never changes at any time or place, and that is independent of a particular human situation. (Von Hayek, 1953: 113–14)

Von Hayek thus adopted a position shared by many social scientists: technology is independent of society and is governed by objective laws in which social choices and representations have no place.

To revert to neo-classical economic discourse, Robbins believed that technology was an external factor, like taste. The former impacted on the functions of production and the latter influenced consumers' preferences. Economists had no more reason to study techniques than to study taste. Both were seen as invariables in a fundamentally static situation.

From 1880 to 1950, neo-classical theory focused primarily on the problem of economic equilibrium and hence on questions of adjustment. Theoretical unity and stability were the conditions required for devising this type of theory. Hence, any uncertainty or change in technology or taste was expelled from the scope of economic analysis. In the 1950s, when economists started to show an interest in the question of growth, technology unexpectedly reappeared.

Neo-classical analysts then tried to explain the origins of the growth of production.[1] They believed that it resulted first from an increase in the quantities of factors of production (capital or labour), following an increase in savings (in the case of capital) or in the size of the population (in the case of labour), for instance. But these increases failed to explain all growth. There was a 'remainder'; production increased even when labour and capital remained constant. Depending on the case, this remainder accounted for between half and three-quarters of all growth. For a long time economists have sought an explanation for this improvement in the efficiency of factors of production, which lies behind most growth. The remainder was explained either by the theoretical or statistical weaknesses of the model, or by the absence of a factor of production.

Since this unknown factor, this 'remainder', had become the main cause explaining growth, neo-classical theory had to bring in an element from outside its field of analysis. Economists tried to incorporate technology into their theory.

Robert Solow was one of the first, in 1957, to take technological change into account in the function of production. He saw production as being determined by three factors: capital, labour and technological developments. Solow's definition of 'technical change' was 'any kind of shift in the production function. Thus slowdowns, speedups, improvement in the education of the labor force and all sorts of things will appear as "technical change"' (Solow, 1957: 312). This is a fine example of the formalism of economic analysis. By referring to technological progress as a 'residual', Solow was able to state that during the 1909–49 period, technological change accounted for 87 per cent of growth (ibid.: 320).

Solow and other economists subsequently tried to incorporate technological progress into capital. Since technological progress changed the shape of the capital goods used, different models could serve to integrate it. They took into account the fact that capital goods were heterogeneous because they incorporated different techniques, and a new technology could be effective only if it was incorporated into a new good. Having eliminated the 'residue' to a large extent, these economists continue to use functions of production that had two factors (capital and labour). In this line of reasoning, the interest in technological progress was only one step in the development of economic analysis leading to a more detailed study of capital goods that took into account their age, in particular.[2] More controversially, David Landes sums up economists' attempt to reduce the residual: 'These efforts did succeed in swelling the proportions assigned to the conventional factors, but it is not clear to me that this is more than a clever bookkeeping device. The reality of knowledge, skill, education, work habits ... remained though now homogenized like cream in milk' (Landes, 1991: 8–9).

If we move on from macroeconomics to microeconomics, technology still seems to be an external element. Mark Blaug clearly defines the dominant position 'The topic considered here is innovations, not inventions: the entrepreneur is viewed as facing a list of known but as yet unexploited inventions from which he may select. How this list is itself drawn up and continuously augmented is an issue that deserves separate treatment' (Blaug, 1963: 13).

This distance between a technical invention and an economic innovation was at the heart of Joseph Schumpeter's work. Schumpeter, considered to be the economist who showed the most interest in technology in the first half of the twentieth century, believed that entrepreneurs were not involved in the innovations that they launched. Their main role was to select the new technical systems that they put on the market. The 'Schumpeterian' entrepreneur was thus a mediator between two impervious worlds. When exceptionally an inventor like Thomas Edison became an entrepreneur, he profoundly changed the nature of his work.

Schumpeter saw innovation as something unusual, unrelated to the continuity of the concept of economic growth. It had several possible forms: 'The launching of a new product or of new forms of organization, the accomplishment of a merger or the opening of new markets' (Schumpeter, 1939: 88). For Schumpeter, innovation was not the follow-up to invention since several of the innovations that he mentioned required no new technical know-how. Yet when he wanted to define innovation with precision, he focused on production: 'innovation is then the fact of establishing a new function of production' (ibid.). In this respect he was part of the neo-classical tradition, following Léon Walras who as early as 1877 defined technical progress as a set of changes in 'the coefficients of manufacturing' (Walras, 1877, cited in Le Bas, 1982: 25). The same definition is also found in the work of researchers like Solow, who specialized in economic growth (see above).

The idea most open to criticism in Schumpeter's position, like that of all the neo-classical economists, is that the entrepreneur taps an endless supply of unexploited inventions. Invention is a rare resource that has a price and for which strict barriers to entry may exist. Firms therefore have to obtain that resource on their own, by investing in research and development (R&D). Even if we agree that the entrepreneur is a mediator between technology and the economy, we have to admit that mediation functions both ways and not only, as Schumpeter said, from technology towards the economy.[3]

Economists of innovation traditionally distinguish between product and process innovation. Although this distinction is often artificial (a new manufacturing process can lead to alterations to a product, or a new product often requires a new manufacturing process), it nevertheless remains meaningful. It allows us to characterize the field covered by most neo-classical economists, who disregard product innovation and focus exclusively on process innovation.

From their point of view, technological progress has no macroeconomic consequences other than creating productivity gains and increasing the supply. Bernard Real (Real, 1990) considers, on the contrary, that along with a supply-side approach to technological progress, we can construct a complementary demand-side approach. He distinguishes between two types of innovation: process and consumer innovations. The former act on the supply, the latter on the demand. Process innovations that induce productivity gains generate growth and employment only if the demand is dynamic. However, the demand can be dynamic only in the presence of consumer innovations. In the post-Second World War years cars, household appliances and televisions were consumer innovations that generated growth. Today that type of innovation seems insufficient to allow development and full employment.

This approach, an extension of the neo-Schumpeterian current, shows that alternatives do exist for integrating research on technological change into economics. On the other hand, in 'standard' economic theory innovation still has little room and the comment made by Henry Bruton in 1956 remains relevant today: 'the lamentable state of our understanding of the origin and process of technical change' constitutes 'the most important deficiency' in contemporary theorizing on economic growth (Bruton, cited in Blaug, 1963: 14).

ECONOMISTS AND THE DIFFUSION OF INNOVATION

The diffusion of innovation is the second focus of economists interested in technological change. The standard model first devised in the 1960s was based on the idea of imitation and was designed to study the spread of innovations. Originally the epidemiological model of propagation through contact was applied. If the progression of an innovation in time is described, an S-curve (slow start, fast growth, then stagnation), said to be sigmoid, is obtained. The main lines of this model are found in studies of internal diffusion throughout firms, entire industries and the economy as a whole. In the first two cases, economists focused only on process innovations.

Edwin Mansfield, one of the first to work on this model, summed it up as follows: 'the probability that a firm will introduce a new technique is an increasing function of the proportion of firms already using it and the profitability of doing so, but a decreasing function of the size of the investment required' (Mansfield, 1961: 762). Zvi Griliches developed a similar approach to account for the diffusion of hybrid corn in the US (Griliches, 1957). One of his conclusions was that since 'differences in profitability are a strong exploratory variable it is not necessary to appeal to differences in personality, education and social environment' (cited in Dixon, 1980: 1451).

In operational terms, the authors of this model tried to construct indicators of speed of diffusion. In a perspective in which a new product will eventually saturate a given environment, the course of its development needs to be foreseen.

This standard model of diffusion can be criticized on many grounds. First, it overlooks the question of the launching of the innovation and, *a fortiori*, that of the conception of the technical object, which has no substance. It is as if the object were developed once and for all before its diffusion, never to be altered again. This finality of the technical object makes it possible to determine the area of diffusion in advance. Behind the model is the idea of imitation, which once again externalizes the technology and cannot explain how the market takes off. This technology–diffusion dichotomy caused the authors of the model not to take into account the subsequent evolution of the technical object.

The model designed to account for a temporal phenomenon (diffusion) is in fact static. It also defines the potential users of the innovation *ex ante* (the upper asymptote of the S-curve) whereas, in reality, the diffusion ceiling changes with time as the technical object develops and other forms of use, never conceived of initially, emerge.

This question of the evolution of the technical object has several dimensions. First, the inventive activity does not come to a grinding halt. Producers carry on improving their product during its diffusion, so that successive generations and different models of the product can be launched. Second, repeated production of the new technical object generates a learning process. US economist Kenneth Arrow was one of the first to study the phenomenon that he called 'learning by doing' (Arrow, 1962: 155–73). In this way, productivity gains are achieved that shorten the time needed to manufacture the new technical object.

Nathan Rosenberg contrasts learning by doing with learning by using (Rosenberg, 1982a). With use, machine operators are able to improve their performance and the human-machine interaction is enhanced. Use also makes it possible to improve the machine, either by the user or by the manufacturer, depending on the size of the alteration.

These different forms of development of the technique undermine one of the underlying hypotheses of the standard model of diffusion: the constant profitability of the adoption of a new technique.[4]

Many economists, notably Rosenberg, have laid the foundations of another economic approach to diffusion. They have rejected the break between invention and diffusion and have taken into consideration the continuity of inventive activity. In their analyses they have taken on board the multiple interactions between suppliers and users, with the repercussions that these have both on use and on profitability. In short, they have constructed a new dynamic and interactive approach. Table 1.1 synthesizes the two economic models of diffusion.

Table 1.1 Two economic models of diffusion

	Standard diffusion model	New theoretical options
Object of diffusion innovation	Set, unchanging entity throughout the process; stable domain of application	Continuity of inventive activity; dynamic conception of the object and of space
Area of diffusion	Characterization reduced to the degree of competition and the size of firms	Coherence and complementarities; pre-existing specific industrial and technological structures
Representation of diffusion	Saturation of an environment by the extension of the innovation from firm to firm, on the basis of a single driver: imitation	Dual movement of saturation of the environment and expansion of the initial field of application
Objective of the model	Speed of diffusion	Explanation of the twofold dynamic of the object and the area of diffusion; analysis of evaluation procedures; evaluation impacts
Status of technical research in the economic theory of diffusion	Exogenous character; explanatory variable of speed of diffusion; technology–diffusion dichotomy	Endogenous character, entangled relations, integration of technology and diffusion

Source: Foray and Le Bas, (1986: 646).

SOCIOLOGISTS AND THE DIFFUSION OF INNOVATION

It is very unusual for different social science disciplines to do empirical research on the same subject. Yet the diffusion of hybrid corn had been studied by two sociologists 15 years before becoming one of the main research subjects of the economics of innovation. For several decades the study by Bryce Ryan and Neal Gross (1943: 15–24), on the diffusion of this crop in Iowa, was a reference among sociologists of innovation. Although the economic and sociological studies were carried out independently, a controversy subsequently broke out between the two teams. In the conclusion to his article, economist Griliches wrote: 'It is my belief that in the long run and cross-sectionally, these variables tend to cancel themselves out, leaving the economic variables as the major determinants of the pattern of technological change' (Griliches, 1957: 522).

Everett Rogers was to lead the battle against the arrogance of the Chicago school economists (of which Griliches was a member). Rogers commented:

> [E]conomic aspects of relative advantage may even be the most important single predictor of rate of adoption. But to argue that economic factors are the sole predictors of rate of adoption is ridiculous. Perhaps if Dr Griliches had ever personally interviewed one of the Midwestern farmers ... he would have understood that farmers are not 100 percent economic men. (Rogers, 1983: 215)

Ryan and Gross had noted in their inquiry that none of the farmers questioned had cited the cost of the initial investment (purchase of seeds) as a hindrance to adoption of the technology. They considered that the economic variable was neutral since it acted in the same way on all the farmers.

This was a typical conflict between economists and sociologists. The former could point out that even if the farmer is not a 100 per cent *homo oeconomicus*, it is highly unlikely that he makes no calculation of profitability when it comes to a production innovation – which the sociologist treats as if it were a consumption innovation. The sociologist would probably reply that the family enterprise in this case does not always have all the accounting tools necessary for economic calculation, and that the choice of the hybrid maize compels the farmer to buy his seed every year. He thus loses a degree of autonomy compared to the traditional way of doing things, where he sows seeds from his own harvest.

Irrespective of who is right and who is wrong, let us revert to the sociological analysis of innovation. Ryan and Gross showed that diffusion functions according to a cumulative process:

> There is no doubt that the behavior of one individual in an interacting population affects the behavior of his fellows. Thus the demonstrated success of hybrid seed on a few farms offers a changed situation to those who have not been so experimental. The very fact of acceptance by one or more farmers offers new stimulus to the remaining ones. (Ryan and Gross, 1943: 23)

The idea of a network of influence thus constituted the core of sociological theory of the diffusion of innovation.

Rogers was the main theoretician of this sociological current. According to him (1983: 124–34) five characteristics determined the adoption of a new technique:

- the relative advantage that can be measured in economic terms but also in terms of social prestige or satisfaction;
- compatibility with the values of the group to which the person belongs;
- innovation complexity
- test possibility (the first users of hybrid corn tested it in a few fields); and
- innovation visibility.

The decision-making process also corresponded to five stages (knowledge, persuasion, decision, implementation and confirmation). Finally, Rogers divided users into five groups (innovators, first users, first majority, second majority and late-comers). These different typologies can be applied to monitor the evolution of the rate of adoption, which – as with economists – is the most important descriptive variable of diffusion and describes an S-curve.

Based on these theories, the 'diffusionist' current often refers to a second study, this time on the diffusion of a new drug. The study was performed in 1957 by Elihu Katz, Herbert Menzel and James Coleman. Katz, who already had a sound reputation as a sociologist of the media, was unaware of Ryan and Gross's work. When he discovered it he drew up a comparative presentation of the two studies.

The first common conclusion was that information on the new product was insufficient to result in adoption. In 1934 (six years after it was put on the market), 90 per cent of Iowa farmers had heard about hybrid corn yet only 20 per cent of them had tried it. It was interpersonal contacts rather than information that led to adoption: 'If the pioneers who adopt an innovation immediately upon its appearance each tell their friends about it, and these friends subsequently tell their friends and so on, the resulting curve of diffusion would look like [the S-curve]' (Coleman et al. 1957; also in Katz, 1971: 772). Phenomena of mutual influence entailed exponential growth (corresponding to the first two parts of the curve). A sociological validity of the epidemiological model common to economists and sociologists was thus identified.

Katz, as a sociologist of the media, was particularly attentive to communication.[5] 'Mass-media' he wrote, 'serve to inform and personal contacts are used to legitimate' (ibid.: 785). He thus granted a key role to opinion leaders who, within the peer group, were to be a motivating force and agents of social change. Rogers adopted this perspective to a large extent. He noted that the first 'adopters' of a new technique were the future opinion leaders, people open to the outside and heavy consumers of the mass media.

The analysis of diffusion curves can also be used in a macro framework. Melvin de Fleur compared the curves of four main mass media: the press, cinema, radio and television (de Fleur, 1970). He noted that even if series of statistics did make it possible to construct S-curves, in the case of the press a diffusion process that had lasted for a century could hardly be explained in terms of mutual influence.

This brings us to a tricky question facing economists: can the theories of micro analysis serve as a basis for macro analysis? As regards diffusion curves, the analysis in terms of opinion leaders clearly cannot constitute the theoretical underpinnings of historical reflection on the diffusion of the media.

At a macro level, the S-curve model gives no indication of the evolution of a medium. The growth may be fast or slow. The saturation level, where the S-curve reaches its asymptote, is sometimes unexpectedly surpassed by far. The producers of the first wirelesses, for instance, never imagined that the same household would buy several sets (Flichy, 1995: 159).

Despite their disagreement on new adopters' motives, economists and sociologists of diffusion have closely related points of view. Both favour the study of diffusion curves and have a unidirectional conception of diffusion in which adopters passively respond to the technological offer, accepting or declining the innovation. The technical object is considered to be a black box that cannot be altered. From the economist's point of view, producers adjust their production systems to the new machines. From the sociologist's point of view, the use environment changes to adopt the innovation. If, however, the difference between the cultural values of potential users and the new technique is too big, the technique is rejected.

The sociologist distinguishes a subset consisting of opinion leaders who have a greater propensity to innovate and to become agents of change among potential users. Unlike the first adopters of the economist, who are motivated by a greater hope of profit, the sociologist's opinion leaders are defined not by their involvement in the new technique but by their openness towards the world outside the professional community (doctors who spend a lot of time in hospitals, farmers who often go into town and so on).

Opinion leaders are thus believed to have a general propensity to innovate and to adopt a new technique, irrespective of what it is. This point is rightly criticized by economists who are surprised that sociologists totally disregard the quality and 'usefulness' of a new product in their analyses. Richard Nelson and Sidney Winter note, for instance, regarding Katz: 'The authors did not even attempt to specify quantitatively the ways in which the new product was superior medically to pre-existing alternatives' (Nelson and Winter, 1982: 269). Moreover, this model is strongly marked by the original context in which it was used, that is, the centralized operation of agricultural modernization and the colonial situation. It seems ill-suited to situations in which technical diffusion is far less linear.

But the main criticism that can be made of the diffusionist model is the fact that it overlooks the technique. We can consider this to be a reflection of a division of intellectual work, and this type of analysis to apply only to the final phase of technological development. This model has inspired very rich empirical studies that describe an entire social network of circulation of an innovation within a society. Yet the theory has a fundamental shortcoming in so far as it refuses to take into account changes to the technical object.

Take the example of the tape recorder or the video cassette recorder (VCR). The reactions of users in the general public prompted manufacturers to alter the characteristics of their machines: simplified use (switch from tapes to cassettes) and limited possibilities in areas of marginal interest to most users (live recordings). A study of the diffusion of these machines ought to include these changes.

Rogers later amended his model and introduced the concept of 're-invention' and the way in which users alter the device they adopt (Rice and Rogers, 1980). Whereas for a long time this behaviour was considered by sociologists of diffusion to be 'noise' in the diffusion process, or an obstacle to the rational and efficient use of a new technique, today 're-invention' appears to be the sign of real integration of innovation into the adopters' culture.[6]

Within the same intellectual approach, Rogers's model was modified and developed by other studies, starting with the threshold models of Mark Granovetter (1978). The 'threshold' is the number of adopters that must be present in an individual network before a person decides to adopt an innovation. The notion of 'critical mass' also appears; this is the point at which enough actors have adopted an innovation for it to succeed. Later, network analysis techniques contributed new ideas and helped to determine who influences whom (Valente, 1995). For network analysts, links among social actors are more important than people's individual characteristics.

THE CLOSED WORLD OF THE TECHNICAL HISTORY OF TECHNOLOGY

Historians' diagnosis has hardly been any different. In 1935, in a special issue of the journal *Annales* devoted to the history of technology, Lucien

Febvre wrote: 'Technology, one of the many words from which history is not made. History of technology: one of the many disciplines still to be created' (Febvre, 1935: 531). On the modern era there does nevertheless exist an abundant literature containing accounts of inventions that Bertrand Gille characterizes so well:

> Regardless of what may have been said, the history of inventions is still largely a mythology and a hagiography. ... Mythology in so far as autonomous and often ill-defined forces are brought into play, and hagiography in so far as the inventor appears to be a character endowed with supernatural powers. And the two are inevitably related since to participate in this mythology one has to have the qualities of a saint, someone in direct contact with the gods. (Gille, 1978: 46)

Rosenberg, US historian of economics, denounced the 'heroic theory of invention' revealed in our everyday behaviours: 'Indeed, not only our patent law but also our history books and even our language cause us to associate a name and a single date with each invention' (Rosenberg, 1982a: 55).

To construct a scientific discourse on technology, it was of course necessary to break away from the heroic theory of invention. A detailed description of techniques and their evolution seemed to be the best way of guaranteeing scientificity. In France,[7] Maurice Daumas's general history of technology is a good example of this technical history of technology which refused to study the political, social and economic context of inventions: 'The main task of the history of technology consists in revealing the reasoning peculiar to the evolution of technology. That evolution has its own internal logic which is clearly distinct from that of the evolution of socio-economic history' (Daumas, 1969: 19). This research approach has an inward-looking perspective. Through a pure effect of statistics, the quantitative increase in the number of technicians leads to a growth of innovation. Daumas wrote:

> The pressure of needs was not enough to accelerate technological progress. A constant growth in the number of technicians was also needed ... This notion of the numeric influence of the protagonists on the pace of technological progress has always been neglected. Yet that is probably precisely where the cause of the spectacular achievements of our era lies. (Daumas, 1962 X)

Daumas's internalist view of technology can be contrasted with that of Febvre, who believed:

> [E]ach era has its technology and each technology has the style of its era. That style shows the extent to which everything in human facts is connected to and influenced by everything else. It shows how technology is influenced by what

we could refer to as general history and at the same time acts on that history. (Febvre, 1935: 533)

It was in this perspective of insertion of technology into general history that Gille set out to write a history 'chained to the material world' (Gille, 1978: IX). He constructed a long period marked by the dominant technologies. Unlike Daumas, the idea was not to write an encyclopaedia of the various techniques, but to analyse technological systems as a coherent set of compatible devices related to one another. The connection between iron and coal in the nineteenth century is a good example of this type of system. In Gille's work, the interdependence of technologies defined both a context and the missing links. This framework he saw as orientating inventors' work. The historian's job was to show its coherence and then to study relations between technological, and economic and social systems.

This wish to structure the diversity of techniques is also found at the heart of an embryonic discipline: 'technology'. In the 1970s, André-Georges Haudricourt wanted to create a new human science at the intersection of history and ethnology, with the aim of establishing a natural classification of objects. His project was akin to that of the biologist wanting to define 'a genealogical classification to account for the real historical relationship' (Haudricourt, 1987: 41). The agricultural tools and techniques on which Haudricourt based most of his work were studied not *per se* but in relation to the gestures that made them work.

Despite their difference, technical historians of technology and technological scientists had the same internalist perspective. They all studied technology as a world apart that had to be described, classified and structured. Once this work had been accomplished, some – like Gille – compared technological and social structures.

In all cases the perspective of technical historians of technology was in phase with that of neo-classical economists. Both Daumas and von Hayek saw technology and economics as two separate worlds, each functioning with a specific rationale.

ALL TECHNOLOGY

As we reach the end of this journey through standard social science theory, which either excludes technology or separates it completely from society, we need to consider another approach that, in a different way, denies technology–society interaction by dissolving the social in technology.

Of the more comprehensive approaches to technological phenomena, one of the most characteristic is that of Jacques Ellul, who believed that there had been a fundamental change in techniques in the late eighteenth and early nineteenth centuries. Whereas there had formerly been only fragmented, locally diffused techniques, the nineteenth-century political and industrial revolution spread their use to all social activity; not only to mechanics but also to law, budgetary rules, economics and so forth. Techniques participated in the process of social rationalization launched by the Enlightenment thinkers:

> This great work of rationalization, unification and clarification is being carried out everywhere, in the establishment of budgetary rules and fiscal organization, in weights and measures and in the planning of roads. That is technology. From this angle, one could say that technology is the translation of men's concern to control things through reason. Make countable that which is subconscious, quantitative that which is qualitative ... (Ellul, 1990: 40).

Ellul saw technology as something far broader than machines: 'The global upsurge of mechanics stemming from the use of energy is subsequent to most of these techniques. It even seems ... that the appearance of various techniques was necessary for machines to evolve' (ibid.). In the final analysis, it was nothing less than a cultural transformation: 'Throughout history, without any exception, techniques have belonged to a civilization where they have been an element surrounded by a host of non-technical activities. Today techniques have encompassed civilization in its entirety' (ibid.: 117).

This technical system that Ellul referred to included all techniques. It was 'indivisible'. Evolution occurred on its own without any outside intervention: 'There is no choice between two technical methods: one inevitably triumphs because its results can be counted, measured and seen, and are unquestionable' (ibid.: 74). The technique imposes itself not only on its producers but also on its users:

> In fact there is absolutely no difference between a technique and its use ... Man is faced with an exclusive choice, that of using the technique as it must be, in accordance with technical rules, or not using it at all; but it is impossible to use it without playing by the technical rules. (ibid: 91)

Any ethical questioning is impossible. Ellul (1990: 91) gives this justification for the atomic bomb: 'Since it was possible it had to be' and concludes: 'That is the watchword of any technical development'.

Hence, the technique becomes completely autonomous; it is no longer controlled or even influenced by a sphere other than social activity: 'The economy can be a means of development, a condition of technological progress or, conversely, it can be an obstacle, but it never determines nor causes nor dominates it. Like political power, an economic system that refuses the technological imperative is condemned' (Ellul, 1977: 153).

This autonomous and uncontrolled technology has become the means to coordinate society:

> The link between the fragmented actions of humans, between their inconsistencies, which coordinates and rationalizes, is no longer human but the internal laws of technology ... the intrinsic unity of technology alone ensures the cohesion between the means and the actions of humans. This reign belongs to it; it is a more insightful blind force than the greatest human intelligence. (Ellul, 1990: 87)

Ellul's thinking had a far greater impact in North America than in France. People tended to misunderstand it in France, especially since in his various writings Ellul often skipped from an all-encompassing sense of the word technology to a far more limited one (a machine). His thesis may seem to be a perfect illustration of technological determinism, but that is not what Ellul's thinking is about. He describes a totalitarian system in which, by its own dynamic, an uncontrolled device devours society as a whole. There is no power to counter that of technology. Every attempt to free humans from technology binds them more firmly, incorporates them more soundly in the technical process. Technology is a form of anaesthetic; it produces *soft* totalitarianism.

The increasing technicization of our society is evident not only in the expanding realm of machines but also in the development of intellectual techniques. Like economics, these are less methods of producing and acquiring knowledge than normative devices. Along with advertising, the technological system spawns its own techniques of legitimization. It favours a form of creativity and a non-conformism needed for its own development.

The totalitarian aspect of technology clearly appears in Ellul's critique of Marxism. He criticizes Marx less for having imagined that the dynamic of productive forces would end up freeing the proletariat, than for having been one of the main actors of the working classes' reconciliation with technology. The bourgeoisie and proletariat are thus united, at least in the same faith in technology!

What can we conclude from this thesis? The question of technology is so all-encompassing that it prevents any particular analysis of technological development. Saying that there is no technological choice since the most effective technique always prevails means either disregarding the complexity of the development process of a technology or having a tautological reasoning where the right technology is always the one that prevails. A technology seems indispensable only with hindsight. More profoundly, in Ellul's thinking there is a logical shift that Cornelius Cartoriadis clearly points out:

> Where one notices that contemporary technological movement has considerable inertia, that it cannot easily be diverted or stopped, that it is extensively materialized in social life, there is a tendency to make technology an absolutely autonomous factor, instead of seeing it as an expression of the orientation of all contemporary society. (Castoriadis, 1992: 125)

Even technological macro systems have been chosen by society. In the final analysis, Ellul's all-technology thesis is as inadequate for accounting for relations between technology and society as the neo-economists' thesis of marginalization of technology.

While we cannot subscribe to theories that disregard technology, relegate it to particular strictly confined areas or subsume it in society, the advantages of studying technology appear clearly at the end of this chapter. Economic growth, industrial or tertiary labour, or mediated communication cannot be analysed without studying technology. It is a stakeholder not only in modes of production but also in lifestyles. Technological innovation thus appears in both production and consumption.

These ideas imply that technology and society have to be conceived of in relation to each other. Technology is not outside economics and society. There is not, on the one hand, a laboratory in which technical objects are produced and, on the other, networks of diffusion. These two phases in the history of innovations have to be articulated in the same theory. The first way of doing so is by examining whether a causal relation or influence exists between the two terms. That will lead us to an examination of the determinism thesis.

NOTES

1. They usually applied a simple function of production such as that of Cobb–Douglas: $Y = A K^\alpha L^\beta$, where Y represents the production for the period, K the stock of capital used, L the quantity of work, and A, α and β constants, such as $\alpha + \beta = 1$.
2. On this question of a remainder, see Maunoury (1972). For an analysis of US growth based on a study of variations of the age of capital stock, see Nelson (1964).
3. For a critique of Schumpeterian positions on innovation, see Maunoury (1968).
4. For a detailed critical analysis of the Mansfield model, see Foray and Le Bas (1986).
5. The links between sociology of communication and sociology of diffusion are multiple since Rogers, a leader of the diffusionist current, subsequently turned his attention to the media. See Rogers (1986).
6. The theories of Rogers and of Katz, amended to take into account interactions between the technological and the social, are currently being revived in French sociology. Dominique Boullier (1989) tried to 'translate' Rogers in terms borrowed from Michel Callon's and Bruno Latour's new sociology of innovation. Opinion leaders become spokespersons, the first step in innovation is likened to the process of *intéressement*, and so on.
7. Although we present primarily French historians, many authors from other countries have written in the same perspective. The best known are Singer et al. (1934–84).

2. Technological determinism – social determinism

Determinism has long been a highly controversial issue in several social science disciplines. The point is to understand how far technology does or does not condition social transformations. Deterministic theories seek to explain a phenomenon in terms of one principal or determining factor. Some scholars think that technology is this factor, others that it is society. Technology itself can exercise causal influence, but technological developments can also produce unintended effects. We study this controversy in different fields of social science. First, economists have wondered whether it is the technological offer or the social demand that drives innovation. Historians of technology have tried to establish whether inventions are inevitable and whether machines make history. Relations between automation and Taylorism have been investigated by sociologists of labour. Finally, for a long time historians and sociologists of communication have debated at length on the effects of the media.

ECONOMISTS' VIEW OF THE ORIGINS OF INNOVATION: TECHNOLOGY PUSH OR MARKET PULL

After analysing eight empirical studies of innovation, James Utterback wrote:

> Market factors appear to be the primary influence on innovation. From 60 to 80 per cent of important innovations in a large number of fields have been in response to market demand and needs ... Innovation also appears to be stimulated by expanding markets and by rising costs of inputs, with innovations aimed at reducing the use of more expensive inputs. (Utterback, 1974: 621)

David Mowery and Nathan Rosenberg (1979: 102–53) rightly criticized this so-called 'demand-driven, market-pull' theory. With obvious glee they listed the contradictions in the seminal studies on which it was based, which focused on commercially successful innovations. The method relied primarily on interviews with company directors. It hardly seems likely that these firms would have invested heavily in the development of an innovation without strong expectations of finding a market. Finally, a critical reading of the founding studies of the 'market-pull' school highlights many data that serve to prove the contrary.

Demonstrating the market-pull theory is therefore tricky. Simply saying that there are potential needs and that evidence of their existence lies in the fact of noting *a posteriori* that a new technique had encountered a market, is stating the obvious. For this theory to be verified, there generally has to be 'a possibility of knowing a priori (before the invention process takes place) the direction in which the market is pulling' (Dosi, 1982: 149). But the technological innovation process develops at its own pace, which is too slow to wait for market signals. Moreover, economists cannot be satisfied with a vague notion of demand. They calculate demand in terms of a mathematical function that links prices to quantities and takes into account consumers' tastes and financial constraints. Although it may be impossible to draw a demand curve in the case of a new product that consumers have not yet been able to assess, the same does not apply to process innovations, which introduce a new mode of production. If this transformation of production techniques results in reduced costs, the supply curve moves without changing the demand for end products. A new supply–demand equilibrium can thus be reached, which explains why economists have always focused on process innovations in their analyses.

Advocates of the theory of the predominance of demand often refer to Jacob Schmookler's *Invention and Economic Growth* (1966). Schmookler showed that inventive activity in an economic sector (which he measured in terms of the number of patents registered in the field) depended on a demand indicator built on investments as well as other data (turnover, value added and so on). This analysis, situated at a macro level, concerned patented rather than commercialized inventions. But we know that there is a wide gap between patents and innovations, and that many innovations have not been patented. In fact, Schmookler overlooked the influence of demand on innovation and focused rather on the way in which the market influenced the allocation of resources for inventive activity – which is obviously not the same thing. Secondary inventions, in the case of a stable technological paradigm, are launched much sooner. We can therefore say that demand influences innovation or, as Michel Zitt put it, that 'the pace of invention is affected more by fluctuations of economic activity than by technological resistances' (Zitt, 1987: 28). On the other hand, when it comes to major innovations and to the appearance of a new paradigm, the resistance of an existing technology is stronger, the gap between invention and innovation increases, and the market-pull theory is to a large extent invalidated (see

Freeman, 1982). Hence, this theory, which is supposed to introduce an automatic link between market trends and technology trends, is unable to explain when and why technological innovation appears.

Theoreticians of the supply-push theory emphasize the essential role of scientific and technological progress. They see no relationship a priori between the advancement of scientific knowledge and demand. Large technical laboratories organize their activities on a long-term basis and are unable to adjust swiftly to changes in market conditions. Finally, certain major innovations seem random and we know that many inventors had no idea of how their findings could be used.

The supply-push theory grants a great deal of importance to the relationship between science and technology. Scientific progress is often presented as an exogenous factor that determines the evolution of techniques since the latter are at the origin of new marketed products. In this perspective, technology is conceived of as simply an application of existing scientific knowledge. Yet many examples exist in which technology preceded scientific work. Engineers provide scientists with measurement tools and sets of empirical facts to analyse. It follows that, in certain cases, engineers even put pressure on scientists to study particular phenomena. If they want to construct their technical object, engineers need to know the laws of the development of the phenomena that they use. In the conclusion to a study on research at the Bell laboratories where the transistor was born, Rosenberg noted: 'Clearly, the main flow of scientific knowledge during this period was from industry to the university' (Rosenberg, 1982b: 155).

The first part of the linear science–technology–market schema therefore seems to be invalidated. What about the last part? The technological supply-push theory has obvious limits, which are symmetrical, to a large extent, to those encountered in the demand-pull theory. The invention–innovation sequence cannot be developed independently of economic factors. In an article on the invention of jet engines, US historian Edward Constant showed that initially a new paradigm was formed independently by the community of engineers, but that in the second stage it was efficiency and cost criteria that determined whether the innovation was put on the market or not (Constant, 1973: 533–72).

A second limit of the supply-push theory is that it assumes a unidirectional organization of the science–technology–market sequence. Although many situations exist in which technology precedes science, there are also cases in which the market is an incentive to innovate. In the final analysis, this theory fails to show when and why the market is created.

Both the technology-push and the market-pull theories therefore seem to be inadequate to explain the birth and development of an innovation. Should we dismiss them without pronouncing ourselves in favour of either of them and consider the debate to be purely academic? Some authors, like Richard Barras, think that each theory has its period of validity:

> In the early phase of major product innovations, technology push pressure resulting from an earlier phase of fundamental research and development is the predominant driving force, whereas in the latter phases of more incremental innovations, the demand pull pressures created by users of the technology become increasingly dominant. (Barras, 1986: 163)

In my opinion that is not a satisfactory explanation. Even if the technological supply is a necessary condition for the development of radical innovation, it is not enough, as the following example illustrates. The year 1948 witnessed the birth (patenting) of the transistor and holography (a system for recording three-dimensional images without a lens). The former was widely diffused whereas 40 years later the latter had still not found a market. Gerhard Mensch, who drew this comparison, wrote: 'the inventor who works in a field outside the prevailing industrial practices and concerns has a sorry fate' (Mensch, 1988: 73). As regards minor innovations that develop within a stable technological paradigm, it is also wrong to think that the demand is decisive. The history of micro-electronic components, very much a case of technology push, attests to this (Malerba, 1985: 283–97). The demand was revealed by remarkable technological potentialities. The sharp increase in integration and reduced costs opened new markets for these components (household appliances, cars and so on).

Even if we agree with Barras that the articulation between supply and demand is not the same in periods in which new paradigms appear and in more stable phases, we cannot say that the supply is predominant in one period and the demand in another. Despite the constant play between the two, the cycles of technological adjustment are different. In a stable period it is quick: technology can respond swiftly to changes in the demand, which, in turn, reacts promptly to technological progress. This adjustment is not always harmonious but it does occur. A case in point is the semiconductor market in which regular crises result from the difficulty of adjusting supply to demand.

In the case of the appearance of new paradigms, the supply–demand adjustment can be quick (for example transistors) or very slow (mechanization of weaving). The uncertainty is far greater in one case than in the other.

The following comment by Christopher Freeman is an appropriate conclusion to this reflection on the respective role of the market and technology in the birth of innovation: 'what makes innovation fascinating is that the market and technology are constantly evolving. As in a kaleidoscope, a sequence of possible new combinations is continuously appearing' (Freeman, 1988: 30). That which is technically impossible today may be possible in a year or two. The most firmly established technology can become obsolete, whereas others that have no market today may find the right niche tomorrow.

Freeman studied a series of innovations by systematically comparing operations that had succeeded with those that had failed. Two important findings emerged from this research project, known as Sappho. First, firms with powerful R&D teams have a competitive advantage over those in the same sector that do not. Second, firms that have in-depth knowledge of their market are in a better position to grasp opportunities afforded by R&D in order to launch new products (Freeman, 1982: 109–15).

These conclusions lead us to a more dialectic conception. Any theory that separates technology and economics, that produces a determinist model in which either one or the other is the explanatory factor of innovation, seems highly disputable.

THE QUESTION OF DETERMINISM IN HISTORY

As we have seen, few historians have shown much interest in technology but some have granted it a pivotal place in social and political developments. After studying an apparently minor technical object, the stirrup, Lynn White showed that it played a key part in medieval history: 'Few inventions have been so simple as the stirrup, but few have had so catalytic an influence on history' (White, 1962: 38). The stirrup, introduced into Europe in the eighth century, enabled the horse rider to be seated more securely: '[It] thus replaced human energy with animal power, and immensely increased the warrior's ability to damage his enemy' (ibid.: 2). The warrior could thus become part of his horse, whose weight enhanced the effectiveness of his weapon. A new military technique was born: fighting on horseback. Men able to master this technique were able to defend their fellow citizens and the new social category of professional warriors appeared. But to maintain horses for war, land was required. By settling men on those lands, feudal lords guaranteed the existence of a mounted army. Suzerains provided their vassals with lands provided they undertook to serve them on horseback. Finally, 'it was the Franks alone who fully grasped the possibilities inherent in the stirrup and created in terms of it a new type of warfare supported by a novel structure of society which we call feudalism' (ibid.: 28).

Thirty years earlier, Richard Lefebvre des Noëttes had written a book on another technique related to horses: harnessing (Lefebvre des Noëttes, 1931). He had shown that a key innovation had appeared in this area in the tenth century: replacement of a neck-strap by a rigid padded collar, which enabled horses to draw much heavier weights. He saw this as one of the causes of the disappearance of slavery. Marc Bloch, who praised the quality of Lefebvre des Noëttes's work, strongly challenged this thesis: 'Causes imply a past history and the decline of slavery did not follow changes in harnessing. It preceded them' (Bloch, 1935b [1963]: 829). Bloch went even further and reversed the hypothesis: 'In the causal relationship that this bold thesis proposes between the phenomena concerned ... it does seem that the link has to be maintained or even extended. And that requires us to switch the terms' (ibid.: 831). Bloch also saw this social causality in a case that he had studied personally: the water mill (Bloch, 1935a: 538–63). This technique, which started to appear before the Christian era, was only really used at the end of Antiquity when slaves were in short supply and other solutions needed to be found to create energy. Later, in the Middle Ages, communal mills replaced domestic grinding because feudal lords forced these practices on peasants and banned domestic tools.

'Invention is not everything', wrote Bloch, 'The community has to accept and spread it. Here, more than ever, the technique is no longer the only master of its own destiny' (Bloch, 1938 [1963]: 837). He cited the case of peasants in the north of France who harvested with sickles even though they were familiar with the scythe. In this way they left long stems that could be gathered for straw, a collective activity that was independent of the limits of the property. Consequently, it was a long time before the scythe, the tool of individualized agriculture, came into general use. In the conclusion to his last article on the subject, Bloch noted:

> There is no main sequence of cause and effect; no order of facts that are decisive everywhere all the time, opposed to perpetual epiphenomena. On the contrary, any society, like any spirit, is the outcome of constant interactions. The true realism in history is knowing that human reality is multiple. (ibid.: 838)

Bloch thus gave up the idea of a social causality for a more complex and multiple causality. The position of Fernand Braudel in *Civilisation matérielle, économie et capitalisme*, is similar: 'In one way or another, advancement or immobility techniques have made up the substance of people's history ... That is why historians who claim to be specialists on the subject are very seldom able to grasp it entirely'. But to this 'everything is technical' he added: 'An innovation is only ever valued in relation to the social pressure supporting and imposing it' (Braudel, 1979: 378). To take an example dear to Braudel, one could say that it was not the rudder that explained the domination of Bruges and its world economy but a conjunction of naval, financial, political and other policies.

Bloch, like Braudel, examined the question of determinism by studying a technique's invention and diffusion. Various US historians have studied the emergence of inventions. Robert Heilbroner considers that different innovations take place in an order set by the laws of nature:

> The steam-mill follows the hand-mill not by chance but because it is the next 'stage' in a technical conquest of nature that follows one and only one grand avenue of advance... One cannot move to the age of the hydroelectric plant before one has mastered the steam-mill, nor to the nuclear power age until one has lived through that of electricity. (Heilbroner, 1994: 241; 55)

Fifty years earlier, William Ogburn had already noted that many inventions had been created simultaneously by several inventors, unknown to one another. He wondered whether inventions were inevitable (Ogburn and Thomas, 1922) and answered that question with the cultural determinism thesis. He used the concept of culture defined by anthropologist Alfred Kroeber, which included both the technological and the social dimensions. In this perspective, culture determines possible inventions to a large extent. Full development of all possibilities is only a question of time, and the specific role of a particular inventor is contingent. Ogburn took the example of electricity, the development of which depended more on cultural elements than on genius (ibid.: 88) and was organized essentially around the telegraphic project. He concluded that technological evolution can follow only one path, for it happens by exhausting most alternatives, so that inventions *are* inevitable.

To this thesis, which S. Colum Gilfillan (1935) adopted, Abbott Usher opposed the idea of 'weak determinism': 'Choices at any given time are limited by the geographical and by the social environment, but over a period of time significant modifications of both the geographical and the social environments are possible' (Usher, 1954: 67). The idea of a spectrum ranging from strong to weak determinism is also developed by Bruce Bimber. On the one hand, technology seems to be *the* driving force of social change; on the other, it can drive social change, only if society provides a favourable environment (Bimber, 1994).

Other authors like Thomas Misa considered, on the contrary, that the determinist thesis appeared only in macro-social research. In studies that adopted a 'micro' perspective, technological causality faded. Misa contrasted Alfred Chandler's work on US firms' vertical integration in the late nineteenth century (Chandler, 1990) with his own study of Carnegie Steel. He criticized Chandler for having built a model from a few case studies and especially for considering that 'the actor's motives are inferred from their actions rather than from their testimony' (Misa, 1994.: 129). Misa, who worked on Carnegie's archives, showed that the firm had no strategic vision of vertical integration but simply the capability for grasping opportunities: 'The challenge for a micro interpretation is to explain the paradoxical result by a non-rational process' (ibid.: 137). In fact, Misa called for historians to 'move between these two levels [macro and micro]' but also to 'take care not to repeat the methodological error of confusing the long-term emergence of patterns at the macro level with the short-term motivations of actors at the micro level'. He suggested that the solution was to 'focus on the meso-level institutions and organizations that mediate between the individual and the cosmos' (ibid.: 140).

Judging by the different studies mentioned in this section, it seems that these two historical hypotheses – technological and cultural determinism – fall on either side of a dividing line that resembles the one identified by economists, between the supporters of technology push and those of market pull. But, as in economics, neither of them stand up to criticism. We need to imagine a weak determinism or interactions between the technological and the social. The conclusion to these economic and historical controversies is that it is necessary to go further than the determinist approach if we want to explain both the birth of a technique and its diffusion. This is what Thomas Hughes proposed: 'A technological system can be both a cause and an effect; it can shape or be shaped by society. As they grow larger and more complex, systems tend to be more shaping of society and less shaped by it' (Hughes, 1994: 112). To take into account these contradictory movements, Hughes introduced the concept of a technological moment ('moment' being used in the same sense as in physics). For him, social determinism served to explain the birth of techniques, while technological determinism explained their maturity.

Although the concept of a technological moment seems more like a collage than a real conceptual breakthrough, at least it has the advantage of reminding us of an important dimension of the question of determinism: the effects of a new technique. This question warrants examination in the fields of communication and labour.

THE QUESTION OF EFFECTS IN SOCIOLOGY AND IN THE HISTORY OF COMMUNICATION

In the introduction to his research on the diffusion of the mass media, Melvin de Fleur noted: 'While most studies of mass communication attempt to unravel ways in which the media influence society, the present analysis tries to bring out ways in which society has influenced the media' (de Fleur, 1970: 59). In functionalist sociology of communication, relations between the media and society are always conceived of in terms of reciprocal influence. Although de Fleur analysed the influence of society on the media, in other works he also studied that of the media on society, like most US functionalist sociologists.

The issue of the effects of the media in the political sphere has received the most attention. The first reflections were based on a very simple schema in which a fragmented public was subjected individually to the influence of the press or radio. In 1944 Paul Lazarsfeld showed, in *The People's Choice*, that interpersonal relations filter the impact of the media (Lazarsfeld et al., 1944). This idea spawned the 'two-step flow' thesis, which Elihu Katz theorized in the mid-1950s and imported into the sociology of diffusion. By multiplying relays of influence, the two-step flow became a sort of multi-step flow.

The question of the effect of the media was addressed from another angle by Marshall McLuhan. His concern was not the influence of a particular programme on the public's behaviour, but the effect of the media as a whole. In his famous aphorism 'the medium is the message' (McLuhan, 1964: 7) we find the idea that each means of communication structures forms of knowledge and, more generally, forms of social organization. Each technical medium solicits our different senses in a specific way. Writing and especially the printed word subject our perceptions to the sense of sight, while oral expression requires simultaneous use of all the senses. Whereas the first era in the history of humanity was that of the tribe and oral tradition, writing induced a 'detribalization'. 'Thus print carries the individuating power of the phonetic alphabet much further than manuscript culture could ever do. Print is the technology of individualism' (McLuhan, 1967: 158). According to McLuhan, this 'Gutenberg galaxy' profoundly transformed society because the book 'created modern markets and the price system' (ibid.: 164). In the twentieth century, the Marconi era was one of retribalization. Even though McLuhan defended the technological determinism thesis, he never tried to show a causal link. Rather, he collected the widest possible variety of citations in a patchwork that constituted the best possible illustration of this electric culture which he heralded.

McLuhan's technological messianism has always been associated with historical theses that owe much to the work of another Toronto academic, Harold Innis, who studied the history of writing (Innis, 1951). In a perspective similar to that of Arnold Toynbee, he studied the rise and fall of empires. According to Innis, the evolution of civilizations is linked to the history of the leading knowledge institutions and to that of inventions in the field of communication. Innis related the spread of papyrus, for example, to the development of the Roman Empire and to bureaucratic power, while the invention of parchment helped to strengthen the clergy and to move knowledge into the monasteries. Later, paper spurred the development of trade in Italy and northern Europe.

Forty years after Innis, similar theses are found in Régis Debray's work. To define mediology, the new discipline that he suggested founding, he adopted a 'matterist' point of view. In his opinion, analysing systems of thinking required the study, above all, of the medias in which writing or images were inscribed: 'The inscription tool depends on the material' [and] alters the spirit of the line but also the features of the spirit of a time (Debray, 1991: 196).[1] Debray's 'matterism' also led him to study the medium of transmission (post and telecommunications). The development of these media tools had to be associated with social phenomena of an entirely different nature. Debray thus associated 'the invention of the electric telegraph' with that of trivial news, 'wireless telegraphy and the birth of "grand" reporting as a literary genre ... the observation satellite and the birth of ecology' (ibid.: 35). But how does this association function? It often appears, as with Innis or McLuhan, in a determinist form. 'Transmission methods and techniques' are presented, for instance, as 'matrixes of mentalities' (ibid.: 52). The media scientist thus tried to shed new light on the subject for historians of mentalities and political scientists. The 'technological determinations of organs of transmission determine more than the conditions of exercise of the hegemony, its content and the organization of struggles for state power' (ibid.: 301). This was the case of communist apparatus in Eastern Europe which had a monopoly on the printed word but was unable to withstand images from the West that no wall could keep out.

Debray's determinism is nevertheless different from that of Lefebvre des Noëttes and Lynn White. The French media scientist considered not that a new medium created a social phenomenon *ex nihilo* but rather that it amplified it and gave it a particular resonance: 'It's a fact that Christianity as a mass movement and the use of the Codex spread simultaneously in the fourth century, without it being clear how much of that triumph can be ascribed to either one or the other' (ibid.: 131). This simultaneity was to transform both religion and the written document. Thus, 'Christianity = Saint John + codex. The codex popularized the word of God and was, in turn, ennobled by it' (ibid.: 132).

The printing press likewise enabled Martin Luther's theses to be widely known. It 'turned the philologist into an agitator and the schoolmaster into an army chief. By conferring on thought an incomparable power of penetration, printing on paper suddenly gave men of letters unprecedented power' (ibid.: 202). This was the effect of the printing press. For both Debray and McLuhan, the determinism of the media amounts to giving those who harness them, almost without their knowledge, the possibility of constituting a dominant thinking.

Historian of literacy François Furet opposed this view of media techniques with the contrasting thesis of a form of cultural determinism. The Reform shattered the authority of scholars, the only ones able to define meaning, by enabling believers to have personal contact with the Bible: 'Luther made necessary what Gutenberg had made possible: by placing writing at the centre of Christian eschatology, the Reform turned a technical invention into a spiritual obligation' (Furet and Ozouf, 1977, Vol. I: 71).

Elizabeth Einstein's perspective contrasts with these two theses of technological and cultural determinism. Although she shared with McLuhan the idea that the move from the manuscript to the printed book profoundly changed knowledge, she did not see transformations in communication as being in any way decisive in social changes in modern times. It was primarily in her method that she differed from the Toronto researchers. For her, there was never any direct influence but rather multiple interactions that occurred very differently, depending on the circumstances.

While most historians have noted the role of the printing press in the development of the Reform, the rapidity with which Luther's theses spread remains a mystery to many of them. 'If we want to dispel it', wrote Eisenstein, 'we should ... move more cautiously, a step at a time, by looking at the activities of the printers, translators and distributors who acted as agents of the change' (Eisenstein, 1993: 152). Eisenstein focused her research on printers, who were at the centre of intellectual activity. They structured the encounters between humanists of different nationalities in contact with them: authors, translators, readers.

But the printing press was also one of the first examples of capitalist enterprise. It maintained relations with the religious and political authorities who patronized or censured intellectual production. The printer was not only a mediator between authors and readers but also an individual whose observation enables us to analyse the multiple interactions among the state, the church, economic activity and the circulation of ideas. Eisenstein's interest in printers is not unrelated to the preoccupations of a current in the social history of art that has studied patrons, their advisers and art merchants, with the certainty of thus being able to interpret relations between art and society in a new way (see Baxandall, 1972 and Alpers, 1983).

With this method, which requires the accumulation of facts and emphasizes the complexity of changes, the revolution of the printing press cannot be summed up in one major effect. The trends witnessed at the time were in fact contradictory. The printing press enabled the Catholic Church to standardize the liturgy and to rigidify its doctrine in the Counter-Reform. But at the same time, owing to reading of the Bible, Protestants became able to practise their faith autonomously. Consider another example, in an entirely different domain: clothing. Through pattern books, clothes became standardized to a certain extent. 'Spanish' fashion spread as far as the Hapsburg Empire, but awareness of diversity became keener. Books were published on costumes in different countries of the world. Engravings also set standards for regional costumes and produced stereotypes. With the printing press a new dialectic of standardization and diversity appeared.

Historical reflection on causality attaches great importance to the question of anteriority. A new medium can impact on a social phenomenon only if it is anterior or simultaneous to it. Eisenstein has shown that more complex phenomena of mutual influence exist. She studied the question of relations between the invention of the printing press and the Renaissance. The fact that the cultural revival launched by Italian scholars and artists started at the time of scribes does not mean that the printing press played no role in the development of the Renaissance. Twice during the Middle Ages, in the Carolingian period and in the twelfth century, identical cultural movements appeared that were unable to alter thinking to any great extent. This was because they were unable to spread to all disciplines and to gather all the scattered pieces of the heritage from Antiquity. The printing press breathed new life into the movement initiated by Petrarch and the Humanists.

The invention of the printing press not only influenced cultural and religious movements in the fifteenth and sixteenth centuries, it also profoundly changed forms of knowledge. Until then there had been no spatio-temporal frame common to men of letters. The manual copying of maps precluded exact reproductions and was the cause of many alterations. Moreover, it was practically impossible to enhance the quality of maps with the comments and discoveries of travellers and merchants. Only the printing press allowed the constitution of a corpus on which consensus could gradually be reached, so that new editions were improvements rather than alterations. The same scholarly work that articulated comparisons with syntheses of the works of Antiquity and added new observations, was necessary if precise historical chronologies and the advancement of astronomy, botany and zoology were to be achieved.

We thus see that Eisenstein's approach was instrumental in reviving studies on the influence of the media. It is regrettable that this type of questioning has not yet been applied to other media such as television. For several decades, sociologists of the mass media wondered whether television influenced elections. Their reflection corresponded to a quasi-behaviourist schema, since the only question was whether TV viewers' behaviour had been altered or not. Few questions were asked on changes in the functioning of politics or the role of parliament subsequent to the advent of television.

THE ISSUE OF DETERMINISM IN THE SOCIOLOGY OF LABOUR

Innis's and McLuhan's theses on the determinism of communication techniques, like that of White on the effects of the discovery of the stirrup, were not unrelated to those of Karl Marx on labour. In *The Poverty of Philosophy*, in respect of a controversy with Pierre-Joseph Proudhon he wrote: 'In acquiring new productive forces men change their mode of production; and in changing their mode of production, in changing the way of earning their living, they change all their social relations. The handmill gives you society with the feudal lord; the steam-mill society with the industrial capitalist' (Marx, 1955: ch. 2, pt 1). This famous citation clearly illustrates the thesis defended by most Marxists, of technical determinism over the organization of labour. In fact Marx's position was more ambiguous than that since a few pages down he wrote: 'The hand-mill presupposes a different division of labor from the steam-mill' (ibid.: ch. 2, pt 2). It is not clear whether the technique precedes the division of labour or vice versa.

This position, with its ambiguities, is a fairly accurate summary of the positions of sociologists of labour a century later, whether they belonged to spheres influenced by Marxism or not.[2] Post-Second World War French sociologists considered that technology determined the organization of labour to a large degree. This was a position peculiar to French research, since during the same period, in the Anglo-Saxon tradition, the influence of labour process theory emphasized managerial strategies.

Many empirical studies have examined the social consequences of technological progress. Serge Moscovici and Georges Barbichon's study of the introduction of cutting machines in coal mines is a fine example of this approach. Miners had specific know-how that enabled them to control collective activity and closely bound their social group. Modernization destroyed this by creating specialization in the use of machines. The specific characteristics of miners' lives tended to disappear, jobs became scarcer and their unions declined. Thus, change 'disintegrated the miner's world. He saw himself as a "reject", an object that was acted on' (Barbichon and Moscovici, 1962: 189). Naturally this change did not take place without crisis or resistance, and the authors proposed a number of measures to facilitate the change. As Norbert Alter noted in a critical review of their work, 'the objective was therefore to adjust men to technological change and economic constraints since the opposite was not conceivable' (Alter, 1983: 67).

These empirical studies of the consequences of technological change were set in the dominant tradition of French sociology of labour, which saw technological progress as a key explanatory variable of industrial organizations. For instance, Georges Friedmann considered technological progress as a fact of civilization that impacted on all contemporary societies. The organization of labour was very much a product of that. Friedmann noted: 'dichotomous organization (separation between thought and the execution of work) is one of the essential characteristics of the technological civilization. It is found in Soviet, Brazilian, Japanese and American industry. ... Taylor went "in the direction of history", for better and for worse' (Friedmann, 1966: 189–90). This position was adopted not only by sociologists but also by workers' unions which, as Danièle Linhart notes, 'did not interfere with the organization of labour since they saw no possible alternative – given the necessity constantly to increase productivity, which they shared with employers and believed was necessary to obtain the comfort to which the working classes aspired and were entitled' (Linhart, 1991: 26).

The consensus reached between employers and unions on technological change led both parties to consider that the division of labour could not be altered. Other sociologists developed a different argument but remained in the same paradigm of technological determinism. Serge Mallet saw the advent of new techniques as a cause of profound social change. He considered that automation created a new class of technicians who were indispensable to the functioning of high-tech industries. These new workers developed specific types of demands concerning the control and management of firms: 'Modern conditions of production now afford objective opportunities for self-management and economic development by those who bear the brunt' (Mallet, 1969: 43).

With each major development in the tools of production – automation in the 1950s and 1960s, computing in the 1970s, office technology in the 1980s – the effects of technological change were studied. However, certain research studies questioned this determinist schema.[3]

Studies undertaken in the UK after the war by the Tavistock Institute of Human Relations,[4] on work in the mines, invalidated Moscovici and Barbichon's analyses. British researchers who belonged to the so-called 'socio-technical systems school', showed that, whenever they had the opportunity, miners reconstructed previous forms of organization of labour based on versatility, despite new automatic machines. In some instances these forms of cooperation, different from those imagined by scientific management theory responsible for the introduction of new technologies, were more economically efficient. We can conclude that while several modes of organization can be used to exploit a particular production technique, the social consequences of technological change are not unavoidable. There are degrees of latitude that workers and experts in the organization of labour can exploit.

Other research based on rigorous international comparisons reached similar conclusions. Duncan Gallie (1978), for example, studied refineries in the UK and France. Whereas the technologies implemented were comparable, the organization of labour as well as hierarchical and professional relations were profoundly different. It therefore seems evident that the technology does not determine the organization of labour. 'Technological determinism does not exist', wrote Claude Durand, 'society uses the technology that it needs for economic and political reasons' (Durand, 1978: 172). 'The organization of labour and of the technology legitimizing it play a part in society that is closely bound to the distribution of power and the exercise of social control. ... The scientific organization of labour is the alibi of the social domination of workers' (ibid.: 179).

The findings of other research on the introduction of computing likewise invalidated the technological determinism thesis. Despite the utopian discourse of computer scientists on the structuring effects of computers in the processing and circulation of data, computerization failed to alter the organization of firms or the division of power in any meaningful way. After studying computer scientists' work, Colette Hoffsaes concluded:

> Due to the inability to know the laws of management, the past is repeated ... Computer scientists want change, but are not able to alter the goals that are known and embodied by the operatives. They will nevertheless help them to do better what they were already doing. They therefore intervene in processes, the stability of which they tend to enhance. (Hoffsaes, 1978: 307)

In the work of Catherine Ballé and Jean-Louis Peaucelle (1972) we also find the idea that, rather than transforming firms, computer technology has simply reproduced the established order. Computer specialists have a highly structured technical culture that can neither be changed nor be an agent of change.

In the final analysis, irrespective of the form and content of a technology, the organization always reproduces itself in the same way. Alter opposed this sociology of reproduction with a sociology of transformation: 'Technological change introduces contradictions, innovations or unexpected effects in a society or firm' (Alter, 1983: 99). In his study of telematics, he showed that there was not only one organizational solution for this new technique but many, and that the best one could be found by trial and error: 'Technological determinism comes into play only secondarily, as a product of interactions and not as the primary factor' (ibid.: 102).

In this perspective, technology becomes dependent on the organization, if not determined by it, irrespective of whether the organization is reproduced without any major change – thus subjecting the technology to its logic, in a sense – or, on the contrary, leaves some areas with leeway that allow the emergence of new organizational solutions based on new technologies.

A new paradigm thus replaced the former model of technological determinism. This new position on relations between technology and society had a far wider diversity of supporters than the former one. They ranged from authors who believed that technology changed nothing in the organization of labour, considering that the same technique could be established with different organizational frames, to researchers who saw machines as a device for materializing social relations, a way of imposing a form of human work. Yet both the old and the new paradigms had the same perspective, the same view of a technique which is already there. In both cases little interest was shown in the birth of the technique and the way in which it became a standard.

If we want to depart from a sociology of effects, we have to try to incorporate into the same analysis the birth of a production technique and its use in the workshop.

David Noble's historical research on numerical control machine tools (Noble, 1984) is clearly set in this perspective:

> For the technological determinist, the story is pretty much told: numerical control leads to industrial concentration and greater managerial control over the production process. The social analyst, having identified the cause, has only to describe the inevitable effects. For the critical observer, however, the problem has merely been defined. This new technology was developed under the auspices of management within the large metalworking firms ... Why did this new technology take the form that it did, a form which seems to have rendered it accessible only to some firms, and why only this technology? Is there any other way to automate machine tools, a technology, for example, which would lend itself less to managerial control? (Noble, 1985: 109–10)

To answer these questions Noble, unlike the sociologists of labour, studied not the firms that used new production machines but the design of the machines themselves. In post-Second World War United States, two industrial alternatives were explored: analogical automatic machines and numerical control machines. The former recorded the outline of a part drawn by a human operator and then mass produced it automatically. By contrast, with the numerical control machine there was no need to memorize human know-how; it was entirely automatic and required complex programming beforehand.

Numerical-control machines had the full support of the US air force which not only financed the Massachusetts Institute of Technology (MIT) to develop these machines, but also strongly encouraged suppliers to acquire them, by providing financial and material assistance. Yet government grants are not enough to explain the success of numerical-control machine tools, which took longer and were more complex to develop than analogical machines. Noble considered it more relevant to examine how the numerical-control machine corresponded to the representations of the corporate planners and managers likely to adopt it:

> [N]umerical control was always more than a technology for cutting metals, especially in the eyes of its MIT designers who knew little about metal-cutting: it was a symbol of the computer age, of mathematical elegance, of power, order, and predictability, of continuous flow, of remote control, of the automatic factory. Record-playback, on the other hand ... retained a vestige of traditional human skills. (Noble, ibid.: 116)

For company managers and production engineers, automation meant no longer having to rely on human judgement that could lead to human errors. But 'the elimination of human error and uncertainty is the engineering expression of capital's attempt to minimize its dependence upon labor by increasing its control over production' (ibid.: 116). Noble concluded: 'The defects of record-playback were conceptual, not technical; the system simply did not meet the needs of the larger firms for managerial control over production' (ibid.: 118). Numerical-control machines allowed that control whereas in the analogical system it was in the hands of the worker operating the machine.

Noble's monograph enabled him to answer the above question on technological determinism: 'If numerical control seems to have led to organizational changes in the factory, changes which enhanced managerial control over production, it is because the technology was chosen, in part, for just that purpose' (ibid.: 120). However, Noble's answer – social relations determine techniques – seems oversimplified compared to the richness of his inquiry. Numerical control tools did not develop only because they met industry's demand, but also because they corresponded to the prevailing technological representations at the time, because the US Air Force used its weight and financial resources to tip the scales, because computer technology was available, and so forth. Thus, even if the hypothesis of a technique determined by social relations seems no more satisfactory than the technological determination thesis presented above, Noble does offer a valuable contribution to the study of production techniques by highlighting the need to study the design and the introduction of such tools simultaneously.

Subsequent to Noble's research, an extensive literature on the implementation and use of numerical control and many other computer workplace technologies was published. In the UK, James Fleck, Juliet Webster and Robin Williams reached similar conclusions: 'it is not technology which brings about the use of Taylorist practices, but rather the particular dynamics and politics of the workplace, based on managerial decisions ... technical change is a political process' (Fleck et al., 1990: 633). Studying specific technologies based on clear-cut design principles, they observed that use did not always correspond to initial predictions. During the implementation phase unexpected problems arose, new solutions were envisaged, controversies developed and, finally, alternative solutions were adopted, different from those initially conceived of.

In France, Michel Freyssenet developed an approach which had a lot in common with that of Noble. In the conclusion to a study on automation, he wrote:

> Productive techniques are sociologically, economically and culturally conditioned in their development and diffusion ... These techniques are also 'constructed' and 'constituted' socially by the objectives, principles, representations and economic and social assumptions in which they are grounded and which are themselves rooted in the employment relationship and in the division of labour that has been related to it for two centuries ... Productive techniques are socially determining because they are socially determined. (Freyssenet, 1992: 494)

In this play of mutual determination, do we need a large, all-encompassing societal frame, the capitalist mode of production so dear to Marxist analysis? Or should we, by contrast – as Freyssenet hints in another part of his text – adopt the anthropological point of view, that 'apprehends productive techniques for what they are, that is, as "social products"' (ibid.: 470). Freyssenet adopted his perspective when studying 'expert systems' used to diagnose machine failure. He investigated the different technical options, the debates between designers and potential users, and those different actors' representations (Freyssenet, 1990b).

Originally the expert systems studied had been designed as tools for reskilling maintenance workers who had to use new machines. In reality, they served primarily to detect failures, and were regularly used by low-skilled workers, without that use enabling them to be considered more skilled. This shows that it is not enough to analyse designers' intentions if we wish to understand the development of a technique. As in the case of expert systems, the technique does not always correspond to the initial project. It is that difference that needs to be explained.

In Freyssenet's explanatory system, his most original study was probably the one on design engineers' culture:

> [That culture] is evident first in the conviction that the technical solution is always more efficient and definitive than another (organizational, social, managerial) for boosting productivity or solving an organizational or social problem. The technique is then seen as the application of scientific laws to a utilitarian domain.

Hence, there can be only one good technical solution, the one which applies these laws correctly and is thereby imposed with the force of truth. The idea of several possible and valid techniques, depending on the goal ... is very hard to accept in this culture.

The vision of a science establishing the laws of nature which mechanically govern it, produces an ideal of technical perfection that does away with unexpected problems and uncertainty. The more a system is closed and reduces human intervention, the more it is supposed to be efficient and perfect. The engineer is therefore expected to foresee everything ... he cannot leave the smooth functioning of the machines that he designs to the user's suspect and inconsistent appreciation. (Freyssenet, 1990a: 11–12)

It is therefore by studying designers' projects in relation to their own and to users' systems of representation, that Freyssenet was able to explain the development of a production technique, from its conception down to its use.

It was precisely this articulation between technology and use that Eric Alsène studied (Alsène, 1990: 321–37). He believed that, as a result of implicit and explicit choices, designers define an 'organizational design' or, more exactly, a 'field of design' out of which it is very difficult to step. When a new technique is introduced onto a production site, the company management can carry out 'organizational redesign'. This is not only a specific adaptation of the organizational model of a particular firm, but also an organizational adjustment that is independent of the technology. The firm uses the opportunity afforded by the arrival of a new technology to implement other reorganizations. Organizational design and redesign therefore constitute the fields in which technology and organization can be harmonized, first in the design phase and then in use. Alsène thus produced a theory of relations between technology and organization that aimed to go further than the opposition between technical and organizational determinism.

Jean-Jacques Chanaron and Jacques Perrin re-examined the technology–organization relationship by bringing in a third factor, science, which they saw as simultaneously determining technologies and the organization of labour. They reconsidered the birth of Taylorism and showed that Taylor wanted to build a scientific system 'based on the study of movements and then on that of time' (Chanaron and Perrin, 1986: 28).

They cited Siegfried Giedion, for whom 'it is the breakdown of movement that is at the heart of the paradigm of mechanization'. According to Giedion, mechanization started with studies of the movements of living beings – walking and running, a bird's flight – and 'really got off the ground with the mechanization of production which is equivalent to dividing human work into the number of operations constituting it' (Giedion, 1983: 306). Hence, there really was a mechanical link in Taylor's theories. 'The scientific organization of work thus seems to be the product of mechanics and the condition of its diffusion. In other words, there appears to be a bi-univocal relation between the mode of organization of work and the prevailing scientific discipline (at a time and in a given space)' (Chanaron and Perrin, 1986: 28). The scientific nature of Taylorism was further reinforced by the fact that, as Michèle Perrot noted, it 'mobilized, even induced all the so-called human sciences' (Perrot, 1979: 495).

Three-quarters of a century after Taylor, new forms of organization of work came to the fore. A new science, informatics, became dominant. Another model for the organization of labour, often described as Post-Fordism, was set in a computer-type of thinking and thereby spread this technology. The various computer-aided production or management systems became the new scientific management. The rules were no longer set by scientific management theory, 'they were incorporated into software. They appeared as objectified in technical constraints in software' (Chanaron and Perrin, 1986: 39).

Thus, mechanics and informatics were not only at the centre of production devices in the nineteenth and twentieth centuries, respectively, but also shaped the systems of organization of labour. In this perspective they were, in a sense, the mediators of relations between technology and the organization of labour.

The idea of a mediation, by a third factor, of relations between technology and organization, was also at the centre of Dominique Monjardet's research. Like many other sociologists, he noted that the same technology could be developed in different organizational contexts. The explanatory element is to be found in market characteristics. In a stable market situation (steady growth) that is captive (no substitute) and monopolistic, 'production managers control the technology in such a way as to impose and justify an internal organization of the firm that guarantees their autonomy *vis-à-vis* the top management and their pre-eminence within the firm' (Monjardet, 1980: 88). This is a form of organization that enhances the value of occupational know-how and is able to manage the mishaps of technical production. In a situation of market instability and stiff competition, firms no longer have to sell what they produce but to produce what customers want. The solution is then to eliminate mishaps from production by automating as much as possible: 'The manufacturer for whom all the elements of the task are predetermined without him having anything to say about it is reduced to a role of surveillance ... and the technical qualification of the job is reduced to a minimum' (ibid.: 90).

Monjardet concluded:

> A and B, the technology and the organization, both have the same status: that of the means available for fulfilling an economic intention in a market: C. Hence,

a relationship between A and B cannot be conceived of independently of C. It follows that any paradigm concerning only relations between technology and the organization is necessarily false. (ibid.: 91)

As we reach the end of this journey among the sociologists of labour, it seems that neither the technological determinism nor the organizational determinism hypothesis is satisfactory. To conceive of the complex articulations between a technology and the organization of labour, we have to study the different mediations that can articulate these two spheres. But, in the research tradition of Noble and Freyssenet, we also have to study the technology from its conception down to its use.

As we have seen throughout this chapter, economists' or sociologists' reflection on the origin of innovation, and that of communication experts or sociologists of labour on the effects of technologies, correspond essentially to a causal schema. Whereas in economics the idea of determination by the supply or the demand is the template, in the other disciplines there has been a gradual shift from a direct causality to an indirect one, that is, to effects of context. Bloch not only inverts Lefebvre des Noëttes's schema, he also weakens the causal principle. The end of slavery did not create the mill but induced its development. Likewise, in Alsène's view technology no longer determines one kind of organization of labour but a 'field of organizational design'. This brings us to the 'weak determinism' model proposed by Usher.

Direct causality or effect of context? In the different disciplines considered, controversies systematically appeared between the proponents of technological determinism and those of social determinism. In economics, the dominant theory of market-pull determination was criticized. In history, the *École des Annales* challenged the technological determinism of preceding work. In the sociology of labour, the technological determinism of the founders of the discipline was likewise challenged by the following generation of researchers. None of these controversies has finally been settled. Therefore, to meet the objective that I have set for myself – to study both the social and the technological components of innovation – I must now put aside the determinist model (whether strong or weak), and turn to constant interactions between technology and society.

As Castoriadis notes,

[A]part from any quarrels on the question of causality in the socio-historical domain, an essential prerequisite of any idea of determination is not met here: the separation of determining and determined terms. It is necessary first to be able to separate the 'technical fact' and another fact of social life, and to define them univocally. It is then necessary to establish bi-univocal relations between the elements of the first class and those of the second. (Castoriadis, 1992: 126)

If we turn away from the issue of causality and undertake socio-technical studies, we can do so over long periods, as Bloch so clearly showed with regard to the mill. Chapters 5 and 6 will be devoted to this question. We also need to analyse the sequence of mediations between technology and society, as Eisenstein did, and examine both technology in-the-making and technology already made, as Noble proposed. My study proceeds through this detailed analysis of the interactions that appear when a technology is still emerging.

NOTES

1. This idea was also developed by Pierre Lévy in *La Machine univers* (1987).
2. For a precise analysis of Marx's position on technological determinism, see Rosenberg, 1982d: 34–53.
3. Note that, at the time, French sociologists of labour had more open positions in some of their writings. For instance, Pierre Naville distinguished productive machines that reinforced a rigid and fragmented organization of labour, from consumer machines that enabled their users to invent new ways of using them, thus allowing them more freedom of action (Naville, 1960).
4. The book by Oscar Ortsmann (1978) contains a clear synthesis of the Tavistock Institute's work.

PART II

A socio-technical approach

3. The anthropology of technology: thinking the technological and the social together

When historians and sociologists want to construct a global approach to technology that highlights its multiple interactions with the social, they often refer back to anthropology. Since Marcel Mauss, anthropologists have known that technology has to be considered as a total social phenomenon. They are aware that it can take on specific characteristics, depending on the culture of a particular ethnic group. They also observe that technological activity has to be related to the ritual and magical activities with which it is often confused. In this chapter I present the way in which anthropologists have studied technology. I then examine in depth the new sociological approaches inspired by anthropology both in the English-speaking world and in France. My study of the French school is more complete because it proposes a very detailed system of analysis of science and techniques. I conclude the chapter with a presentation of research on the use of contemporary technical objects.

CULTURAL TECHNOLOGY

French ethnology has focused on the material culture of the populations that it has studied. It has tried to define this research by calling it 'cultural technology'.[1] By choosing this term it has specified its project in relation to that of other researchers presented in the preceding chapters, who wanted to construct a 'plain' technology. André Leroi-Gourhan, the main representative of this school, wrote:

> The continuity between the two sides (technical and social) of the existence of [human] groups has been insightfully expressed by the best sociologists, yet as an overflowing of the social onto the material rather than as a two-way current whose profound motivation is material. Accordingly, we are more familiar with glamorous interaction than with daily interaction, with ritual services than with trivial ones, with the circulation of dowries than with vegetables, and far more

familiar with the thinking of societies than with their bodies. (Leroi-Gourhan, 1964, I: 210)

The question of the body is central in the 'cultural technology' approach. Mauss laid much emphasis on techniques of the body, on the postures of daily life (Mauss, 1967: 30–32). Leroi-Gourhan not only studied the material life of societies, but he also related techniques to bodies. For instance, he noted that one cannot study chairs without studying ways of sitting and of resting (Leroi-Gourhan, 1950b). He also explored the relationship between tools and gestures: 'The tool exists only in an operative cycle. It attests to it faithfully because generally it carries significant traces, but in the same way that a horse's skeleton carries the mark of a herbivorous being that can run fast ... The tool only really exists in the gesture that makes it technically effective' (Leroi-Gourhan, 1965, II: 35).

Leroi-Gourhan, who was also an archaeologist, set this articulation of the tool to the body into what can be termed a paleontology of the technical gesture: 'The Australanthrope seems to have possessed these tools like claws. He seems to have acquired them not through a sort of flash of genius that caused him one day to pick up a sharp stone to arm his fist, but as if his brain and his body gradually emitted them' (Leroi-Gourhan, 1964, I: 151). He considered that nothing in gestures marked the definitive break between this primate's teeth and nails, and the tools of the earliest humans. Hence, there was a biological evolution: 'The manipulatory action of primates, in which gesture and tool merge, is followed with the first humans by that of the hand in direct motivity, in which the manual tool becomes separable from the gesture' (Leroi-Gourhan, 1965, II: 41).

Leroi-Gourhan's reflection on tools and gestures corresponded to his view of technology which he saw as a series of operations. In another work (Leroi-Gourhan, 1950a) he emphasized the study of operative sequence. He described both the series of technical gestures required by a manufacturing process and the different steps in an agricultural cycle or in the processing of an object from the raw material stage down to the end product. Another concept that also played an important part in this book was the technical whole. Several technical groups could thus be characterized, all of which were based on the same mechanical or physical principle: 'For instance, when we have the principle of the wheel, we can also have the tank, the potter's wheel, the spinning wheel, the wood lathe, etc. When one knows how to channel compressed air, one can have a blowpipe or a syringe' (Leroi-Gourhan, 1950b: 39). This concept could be used to analyse the interaction between modes of thinking and technical facts. More generally, Leroi-Gourhan considered that technology was spawned by interaction between what he called the *'milieu intérieur'*, that is, what ethnologists would call the culture of a human group, and the *'milieu extérieur'*, that is, the natural environment. The latter imposes a number of universal constraints. Generally speaking, all tools comply with the laws of geometry or rational mechanics. For example, axes are given handles and arrows are balanced on a third of their length. In this respect Leroi-Gourhan referred to 'technical convergence'. Each ethnic group's culture influences the character of its tools. It is not simply a matter of secondary or decorative elements, but also of specific arrangements that concern complex objects such as the forge or the plough.

Within an ethnic group's culture (or *milieu intérieur*), Leroi-Gourhan distinguished the technical environment (*milieu technique*), that is, the stock of techniques and their uses. This technical environment evolves by borrowing inventions. It is therefore very difficult to separate these two forms of evolution in practice: 'This arbitrary distinction ... is the source of the greatest difficulties in ethnology, of the impossibility in which the observer finds him/herself almost as a matter of course, to opt for either technical convergence or borrowing' (Leroi-Gourhan, 1950b: 420). Leroi-Gourhan thus refused to enter into one of the main debates of the founders of ethnology: do techniques spread through successive borrowing or concomitant inventions? We could consider, like him, that this debate is pointless because in the final analysis a technique is meaningful only when it is used.

It is unquestionably in situations of technological change that the interactions between technology and culture are easiest to analyse and that such analysis is the most fruitful. Leroi-Gourhan studied reindeer importing by Inuits in Alaska. This development of breeding was promoted by the US government in the early twentieth century to provide a new source of food, after the disappearance of whales, which had been decimated by fishing companies and the gold rush. The switch from hunting and fishing to breeding changed not only the Inuits' technical environment but also their culture as a whole. Their traditional nomadic lifestyle disappeared as they adopted a sedentary habitat inland. More profoundly, their entire social and religious organization was altered: 'This disruption was accentuated by the increasingly complete exclusion of raw materials from the sea. ... The laborious system of land–sea, male–female coordination and the prohibitions that formed the backbone of the moral order totally disintegrated' (Leroi-Gourhan,1950b: 390; see also Leroi-Gourhan,1936).

This monograph, which highlighted the importance of the coherence of the cultural system, was consistent with the theses of US anthropologist Melville Herskovits, who considered that a culture was structured around a strong point that he called the 'cultural focus'. He took the example of the Toda in India, who had centred all their economic activity around care for their buffalo and the milking cycle. All social and religious life revolved around the buffalo. In the early twentieth century the British army destroyed the sacred dairies of a Toda village to build a shooting range. The Toda, whose entire cultural system was thus de-structured, subsequently gave up their buffalo to breed cattle and grow potatoes (Herskovits, 1948: 542).

Nevertheless, these monographs remain exceptions, as Leroi-Gourhan acknowledged when he wrote in the late 1960s:

> The study of technical activities remains one of the areas that needs to be explored as a matter of urgency. ... The precise description of facts relative to techniques must be accompanied by research on all the connections that make the studied organization an entity. In reality, technical unity in time and space lies neither in objects nor in institutions but in relations. (Leroi-Gourhan, 1968: 1820–21)

Unfortunately this programme has never really been established. The main activity of Leroi-Gourhan's school was less to study the interaction of technology and culture than to construct systems of classification in the framework of an evolutionary theory. Leroi-Gourhan remained marked by the taxonomic approach of botanists or paleontologists: 'Technological evolution in its most lofty forms is no different from evolution as biology designed it' (Leroi-Gourhan, 1950b: 361).

Robert Cresswell, one of the main representatives of the contemporary school of ethnology of technology, critically reviewed the work of this current: 'Too many detailed descriptions, although balanced and useful, overlook the social relations of the context in which the tools or objects are set. ... It is *objects* rather than *processes* that have been the main focus of researchers in cultural technology' (Cresswell, 1983: 143–4, original emphasis). What a bitter conclusion for researchers who have always claimed to do the opposite!

This shortcoming, which could be observed in many studies published by the journal *Technique et culture*, was not peculiar to the French school only. It also existed in the work of British anthropologists of 'material culture'. One of them, Barrie Reynolds, commented:

> Within any society no item of material culture stands in total isolation from other material phenomena nor from the human members of that society, their beliefs and their behavior. The interaction between the items and the other elements is a continuing process. Even where an item is removed from its original setting, for example to a distant museum, this process continues though within the new setting. What we are seeing is a network or system of interaction of which the material item forms the core. (Reynolds, 1983: 213)

Reynolds then went on to propose studying this object – which, in a sense, moved its socio-technical network around with it – in a museum, something which seems to be in total contradiction to his definition of material culture.

Large collections of technical objects in ethnographic museums have also prompted certain researchers to analyse technology as a language or at least as a system of signs (see, for example, Laughlin, 1983: 9–29). Until now these 'structural technology' projects do not seem to have resulted in relevant studies.

Is the anthropology of technology a dead-end? It reiterates the need to study interactions between technology and culture, yet contents itself with a technological anthropology of technology. The grand evolutionary or structuralist frescoes of technology seem so complex to build that one can hardly imagine them including cultural phenomena. While monographic studies may seem more suited to highlighting technology–culture interactions, one has to admit that many of them are disappointing in this respect. However, this is due less to their authors' inability to meet their objectives, than to an error in the choice of a subject of analysis. In a situation of stability, relations between technology and culture are difficult to highlight, since technology seems so perfectly adapted to the society in question. But when analysts study situations of change, they see interactions 'in-the-making'. In research on salt extraction in Niger in the pre-colonial period, Pierre Gouletquer shows that the sources of this new activity lay in profound change:

> The invention was not the appearance of a technique used to extract salt from sludge. The invention, so to speak, lay in the political and economic restructuring that made it possible to take commercial advantage of a potential formerly neglected. ... In other words, the faculty for innovation cannot be analysed in an isolated technical process. The driver of change lies in a redistribution of political powers and a restructuring of territories on the scale of an entire climatic zone; problems of diffusion or invention of techniques become secondary. (Gouletquer, 1991: 347–8)

Colonial situations, with the violent culture shock inherent in them, are another interesting case of change in which the interactions between technology and society appear clearly (in this respect the example of the Inuits and the Toda cited above are exemplary). Today, situations of technology transfer in developing countries can also be fruitful objects of analysis, as a major research tradition on the subject attests.[2] Innovation likewise has much to teach us, especially if we investigate more than simply the construction of a new material device and include the whole range of its uses. Certain researchers in communication are sometimes criticized for focusing only on new technologies and disregarding the 'old' ones which actually comprise the core of the mass media today. Even if that choice is often made by their sponsors rather than by the researchers themselves, I believe that they have opted for a particularly fecund area for constructing a new anthropology of technology (see Jouët, 1994: 73–90).

TECHNOLOGY AS A SOCIAL ARTIFACT

Even if the ethnology of technology and especially the 'cultural technology' currents have not been as fruitful as the seminal works cited above allowed us to expect, some case studies do show the interest of an approach that rejects the separation between technology and society, and not only links the object to the gesture but also resets it in the entire process of its construction and use. Ethnological monographs point to a particularly rich area for the study of technology: periods of change and innovation. It is nevertheless in other studies, those that the new Anglo-Saxon sociology of science has produced, that we find a far more profound analysis of the various interactions of science and technology in-the-making.

The grand model of the epistemology of science, as developed until the 1950s, was based on the idea that scientists developed an autonomous system of thinking related to a specific method. Thomas Kuhn's theses on scientific paradigms (Kuhn, 1962) were the first to undermine this model. Kuhn was one of the pioneers in organizing the study of science around scientific controversies, by articulating their cognitive and social structure. He showed that in a paradigm the cognitive and the social are inseparable. Apart from his thesis on scientific revolutions, he also described the daily activities of researchers as part of his reflection on 'normal science'.

In the early 1970s David Bloor defined the new rules of methods for inquiry into science that became known as 'strong programmes'.[3] He noted that social explanations of science were always asymmetrical (Bloor, 1976). Previous accounts of the development of science were always written in the light of hindsight beliefs as to which hypotheses would eventually be considered true. Since what is today held to be true may well be refuted by subsequent scientific developments, Bloor proposed an impartial attitude regarding beliefs considered to be false. He advocated a principle of symmetry, whereby the same explanatory factors would be used for 'true' and 'false' theories. Apart from this reflection on method, another British researcher, sociologist Harry Collins, proposed a detailed study of 'science in-the-making' (Collins, 1985). He challenged the principle of replication of experiments and showed that it was more often recommended in manuals than applied in laboratories. Collins also highlighted the importance of negotiation between researchers for defining phenomena and establishing results.

In subsequent years a number of historical studies were published by historians of science, on various issues of scientific controversy (Shapin and Schaffer, 1985). The authors analysed these debates without taking into account their outcome. Their aim was to make the thesis that was eventually rejected seem plausible and rational, and thus to show that the same scientific fact could be interpreted in several ways. Shapin and Schaffer had realized that the trade-off between different interpretations was based not on the observed reality but on the resolution of a controversy between different theses, which they analysed essentially as a social mechanism. The question of truth, one of the most fundamental in the epistemology of science, therefore made way to an approach that sought to analyse how consensus was gradually reached in the scientific community.

The new English school of sociology and history of science that was thus formed, studied controversies as revealers of the flexibility of scientific interpretations. These researchers nevertheless considered the final explanation for the resolution of these conflicts to be social: through the play of controversies that mobilize many non-scientific elements, science is socially constructed.

It was this thesis of social constructivism that English scholars have applied to technology through the Social Shaping of Technology (SST) perspective (Williams and Edge, 1996), and the Social Construction of Technology (SCOT) approach developed by Trevor Pinch and Wiebe Bijker (Pinch and Bijker, 1987).[4] In this way, a multidirectional model was built that contrasted with the traditional unilinear one. In the latter, the passage from basic research to the market comprises several stages: applied research, technology development, and production. In the multidirectional schema, by contrast, technology evolves simultaneously in several directions.

Pinch and Bijker showed that for each technical artifact one could determine the 'technological frameworks' constituting the social and cognitive environment within which manufacturers and users designed and used the same object. They also distinguished different 'relevant social groups' of actors involved in the process of creation or in the definition of uses. Even if a social group may have its own technological framework of reference in certain cases, in others this framework is common to several groups of actors.

Each group concerned by an artifact defines the problems that it has encountered and the solutions that it would like to see implemented. This is when a technological controversy appears that will be solved by social mechanisms of conflict and negotiation. To define the relevant social groups for their analyses, sociologists have to start with representations of the technical object: 'The key requirement is that all members of a certain social group share the same set of meanings, attached to a specific artifact' (ibid.: 30). In relation to their representations, these social groups will socially construct the technical object through the definition of problems and their resolution.

The question of closure of a dispute is therefore central in the SCOT model: 'Closure in technology involves the stabilization of an artifact and the "disappearance" of problems. To solve a technological controversy the problems need to be solved in the common sense of the word. The key point is whether the relevant social groups see the problem as being solved' (ibid.: 44). This closure is an end. Once achieved, it is seen as being very difficult to change.[5] Pinch and Bijker distinguished two ways of resolving a controversy: either purely theoretically, or by shifting the problem. They applied their model to the case of the bicycle in the late nineteenth century and showed that the controversy over safety was finally resolved by the effects of rhetoric. The manufacturers of a new model, the 'easy' bicycle, claimed to have all the advantages of other models as well as being totally safe. Another controversy concerned inflatable tyres. The advocates of this solution claimed that the tyres would absorb the vibrations of the wheels, while its opponents saw them as an unsightly accessory that spoiled the bicycle's appearance. The former imposed their system by showing that it solved another problem: speed. Controversies can be solved by moral arguments as well as technical ones, and several bicycle manufacturers developed models that could be ridden by women in long skirts. The idea was to enable women to ride a bicycle astride. This issue could of course also have been solved by simply agreeing that women could wear trousers.

Thus, the social constructivism approach was not content to assert that the form of technical objects was influenced by society; it considered all objects to be social: 'The key point is not that the social is given any status *behind* the natural; rather, it is claimed that there is nothing but the social: socially constructed natural phenomena, socially constructed artifacts, and so on' (ibid.: 109, original emphasis). It is clearly in this sociologism that the originality of Pinch and Bijker's thesis lies, for the idea that science and technology are constructed and not simply a recording of the results of experiments is accepted by most of the theoreticians of science and technology. Some of them, who can be situated in a neo-Kantian tradition, explain this construction of scientific facts in terms of symbolic systems: it is the underlying structures of thinking that determine the frameworks of observation and interpretation. The social constructivism thesis contrasts directly with this tradition, represented by Gaston Bachelard, in particular.

Hence, Pinch and Bijker departed from the classical approaches to technology studied in the first two chapters, in their refusal of the unilinear model and their preference for a multilinear one, which seems to correspond better to what any observer of technological innovations sees. But the most innovative element in their theory was unquestionably the fact of moving away from the traditional break between design and use, production and the market.

Consider this point more closely. Louis Quéré judiciously criticizes the classical technology/use distinction:

> What is remarkable about this type of approach is that it treats everything preceding diffusion itself like a black box, is incapable of granting it an *internal sociality*, and considers the diffused object or process as a pure singularity devoid of any generality. The *montée en généralité*, that is, the shift from the particular to the general, is conceived of only in terms of diffusion, as an external addition and not an internal component of the development of a new product, machine or process. It is in this respect that this classical approach has the same shortcoming as epistemological discourse. In the study of innovation it removes an irreducible core from sociological investigation, due to its inability – as a result of its implicit theory – to recognize that it has a social character. It places a black box at the heart of any innovation; only that which comes before or after can be explained in standard sociological terms. Regarding the phase prior to an innovation, the social relations or cultural models incorporated in the new entity are highlighted, as are the choices from which it results, the social logics that it comprised, the ideology of the engineers and technicians who designed it, etc. As regards the post-innovation phase, there is the whole range of concepts and distinctions of diffusion studies. (Quéré, 1989: 103i added emphasis)

Quéré clearly defined the project of a new sociology of technology which corresponded to that of Pinch and Bijker. Their theory was an attempt to 'confer an internal sociality' on the technical artifact and to make generalization an internal component of the development of the new artifact. They saw the technical object as being constructed by social groups of designers and users through negotiations, subsequent to controversies and prior to problem solving. Unlike the approach of a certain sociology of the uses of new technologies, it was not simply a matter of a transformation resulting from users' reappropriation of a technical object, but a real attempt to open the black box and to discover 'the social link in the machine', as Quéré put it. However, although Pinch and Bijker's project seemed original and really fruitful, its implementation could often be criticized. Their notion of a social group was extremely vague. Even if, in certain cases, the term referred to groups with a specific identity, in others, and especially for user groups, those groups existed only in virtual terms for they had neither identity nor means of expressing an expectation as regards the technology. In their study of the bicycle, the authors distinguished sports cyclists, bicycle tourists, women cyclists who could not find bicycles suited to their clothing, and even anti-cyclists who put spokes in the wheels of the 'little queen's' fans. These groups may have had a common representation of the bicycle but did not all have the same means to act in situations of controversy. Sportsmen were organized in groups and met at races, while bicycle tourists were not. Finally, since use of the bicycle by women was socially condemned, this group existed only in theory. Pinch and Bijker defined it as such: 'Engineers and producers anticipated the importance of women as potential bicyclists' (Pinch and Bijker, 1987: 34).

This comment highlights the weak point of Pinch and Bijker's model. Their reference was that of scientific controversy, where producers and consumers of knowledge belonged to the same community. There was no real distinction between the two and both participated in controversies. In the technical domain the situation is different. The break between producers and users is much deeper. In some cases users can be stakeholders in negotiations to define the artifact.[6] In others their presence is virtual; they are involved in the debate only indirectly, via the representation that other groups – here, manufacturers – have of their position. Langdon Winner's question, 'Who says what are relevant social groups and social interests?' (Winner, 1993: 369) is indeed relevant.[7]

In view of this, can this question of confrontation between social groups be placed at the centre of the analysis? Should we not also examine the way in which designers (who, in spite of everything, are still the main actors in the construction of the technical object) represent users in various social groups and the way in which they might react to a particular technical change? User representations are a key element mobilized in controversies between designers. Although including design and use in the same analysis of technology may seem essential, we should not postulate a false symmetry between the two. Designers and users do not participate in the same way in the social construction of the technical object. The former incorporate uses in their machines as a social link, but social uses are present in the laboratory only virtually. Users experience the machine as a social link, but their creativity as users is shaped by the constraint of resistance to a technique with limited functions and possibilities. Although socio-technical negotiation exists in design and use, it is of a very specific nature in each case. Designers incorporate virtual uses into the machine if it is an entirely new device or real uses if it is based on an existing machine. Either way, they are simply representations of uses. Users, on the other hand, negotiate with a machine that offers them a use potential which they can alter but which also has its own resistance. Hence, there is negotiation either with representations or with an object. Social representations and technical objects are both malleable and set, but in different ways.

Pinch and Bijker's approach integrates reflection on design and use by conferring an internal sociality on the technical object. Yet its weakness is its sociologism: it reduces technological conflicts to social ones. Just as the cultural technology approach tended to favour the technical object over the social process, so social constructivism reduces technological choices to social ones. In the association of technology with society, one of the two always prevails. To solve this dilemma we need to consider another school of thought, the French sociology of translation, which has eliminated the hierarchy between technology and society, and considered the two as being redefined and reconstructed simultaneously.

TECHNOSCIENCE AND NETWORKS

As in the English-speaking world, the new approach was born in the sociology of science. Initially it grew out of Bruno Latour's ethnographic project for observing 'laboratory life' from a radically new angle, without any assumptions on the content of scientific activity. In 1979, Latour and a British colleague Steve Woolgar published *Laboratory Life: The Construction of Scientific Facts* (Latour and Woolgar, 1979), which articulated their approach *vis-à-vis* that of Collins, presented above. Latour and Woolgar devoted much of their work to the production and circulation of scientific articles. They analysed science in-the-making as a rhetorical activity: above all, the researcher has to be convincing. Scientific facts are not inscribed eternally in nature, where the scientist simply has to discover and reveal them. On the contrary, the researcher has to construct facts, to extract some islands of order from the disorder characterizing nature.[8] From the different measuring tools used by researchers, only noise emerges (in the sense of information theory). It is up to them to extract the salient features, to construct new scientific statements. A new statement will become a scientific fact only if the researcher successfully convinces colleagues, and ensures that their reply is not 'But you could also say that ...' as they propose equally probable statements. To construct and impose a fact, researchers have to use all the resources of rhetoric, and to put the phenomenon under study through different trials selected to disarm critics. These trials, which correspond to various tests and measurements, require the infrastructure of a laboratory.

Experts in science tell us that to ensure successful trials, to consolidate facts in-the-making and to combine elements of various origins, we need the scholar's inventiveness and genius. Latour and Woolgar showed, on the contrary, that flashes of genius are usually explained by fortuitous circumstances, chance encounters, conversations with colleagues. The explanation in terms of a new idea is most often reconstructed afterwards by the scientist him- or herself or by historians. But when the ethnographer happens to be there to observe science in-the-making, he or she can see how much flashes of genius owe to circumstances. After a while the unquestioned scientific statement is reified, incorporated into the scientific knowledge of the day, and becomes a 'black box', a stable scientific fact considered by everyone to be part of nature.

Latour subsequently systematized this initial model for analysing science and extended it to the study of technology. His collaboration with Michel Callon spawned a sociological theory of technoscience developed at the Centre for the Sociology of Innovation (CSI) of the École des Mines de Paris. The associationist principle was enhanced and expanded, and the key metaphor became the network whose ramifications were described as spreading further and further to include not only humans but also 'non-humans'. The other key element in Callon and Latour's work was their rejection of any separation between technoscience and society. Science, they said, was not a social construction but a network connecting different elements of socio-nature.

In this theory, developed by Latour in *Science in Action* (Latour, 1987) and by Callon in *La Science et ses réseaux* (Callon, 1989a), our two authors used another metaphor, that of the spokesperson. They borrowed the fictive character of the Leviathan from Thomas Hobbes's political philosophy (Callon and Latour, 1981). Like the sovereign in Hobbes's work, who is authorized by contract to talk on behalf of the people, Callon and Latour's spokesperson can condense the interactions developed by a host of actors. Thus, a macro actor can be substituted for a heterogeneous mass of micro actors. But by switching from the field of political philosophy to that of science, the notion of a spokesperson becomes much more vague. The spokesperson can represent a laboratory or institution, in which case the question of its representativeness and legitimacy arises. It may also be the spokesperson of microbes or electrons: 'The author behaves as if he or she were the mouthpiece of what is inscribed on the window of the instrument' (Latour, 1987: 71), wrote Latour. He compared this author with the trade union representative surrounded by a mass of strikers: 'The result of such a set-up is that it seems as though the mouthpiece does not "really talk", but that he or she is just commenting on what you yourself directly see, "simply" providing you with the words you would have used anyway' (ibid.: 73). Even if, in both cases, the presence of humans or non-humans around the spokesperson precludes any cheating or abuse of the situation, there is nevertheless a fundamental difference between the two that Latour failed to mention: the researcher chooses the non-humans that he or she wants to use, whereas the trade union representative is chosen by the humans.

The third associationist metaphor used by Callon and Latour was that of translation, borrowed from Serres. Latour defined it as follows: 'In addition to its linguistic meaning (relating versions in one language to versions in another one) it also has a geometric meaning (moving from one place to another). Translating interest means at once offering new interpretations of these interests and channelling people in different directions' (ibid.: 117). Of the many examples, consider two cited by Latour and Callon. The first was a letter written by Louis Pasteur to a cabinet minister to request a 2,500 franc grant to study the fermentation of wine. Pasteur justified his request by the fact that agriculture was one of France's main riches, and that its importance had increased following the free-trade treaty with England.

The second example was that of a fuel cell which the French electricity utility, Électricité de France (EDF), wanted to install on the future electric vehicle. To develop these cells, researchers considered that they had to have efficient electrodes and therefore decided to study a particular type of electrode (Callon, 1989b: 179–82). These two examples are perfect illustrations of the heterogeneous nature of the concept of translation. In one instance it is a matter of pure rhetoric: referring to 'the grandeur of France' to obtain a grant; in the other, a question of segmentation of a technical problem. Of the various questions raised, concerning the fuel cell, that of its functioning which could be observed with the new type of electrode seemed to be the most important. Even if, like Latour, we willingly agree that obtaining funds is as much part of a scientist's job as laboratory work, and that institutional negotiations obviously have consequences on scientific choices, the rhetoric of persuasion of the financier seems fundamentally different from that of electrons.

Translation, spokesperson: these two concepts overlap to a large degree[9] and relate to a sort of generalized associationism. 'What then is a sociologist?' ask Callon and Latour, and answer with: 'someone who studies associations and dissociations, that is all' (Callon and Latour, 1981: 300). The key concept of their theory is therefore the network. A scientific fact or a technical device comprises an assemblage of heterogeneous elements, articulated in a network. The technical (or scientific) work consists in strengthening that network to make it more durable.

The network is composed of both physical elements (electrons, winds, tides and so on) and social actors. In fact there is no reason to distinguish the technological from the social, especially since boundaries are fuzzy and vary from one author to the next: 'Innovation is the art of interesting a growing number of allies who make you increasingly strong' (Akrich et al., 1988: 16). To reinforce their networks, engineers use all available means; they enrol both human and non-human actors. The success (or failure) of an invention stems not from the quality or appropriateness of a technical device nor the capability of the innovation to meet a social demand; everything depends on the soundness of the network: 'The difficulty of innovation is measured by the fact that it assembles in the same place and in the same combination a group of interested people, a good half of whom are prepared to desert, and a set of things, most of which are likely to break down' (Latour, 1992b: 56).

The radicality of Callon and Latour's point of view led them to reject the 'great divide' separating society from science and technology in our modern world. They likewise refused to grant the sociologist a position external to technoscience; in other words, they blurred the boundary between technological research and sociology. The engineers who elaborate new technologies, wrote Callon, 'construct hypotheses and forms of arguments that pull these participants into the field of sociological analysis. Whether they want to or not, they are transformed into sociologists, or what I call engineer-sociologists' (Callon, 1987: 83). Callon illustrated this with the example of the controversy between EDF and Renault in the 1960s, over the electric vehicle. The EDF engineers adopted a view of French society that had a lot in common with the one developed in the same period by Alain Touraine on the post-industrial society, in terms of which technocratic decision makers had to rely on the social movement opposed to the motor car to propose another form of less-polluting transport. To counter EDF's electric vehicle projects, which would eventually entail the disappearance of Renault, the company's engineers not only improved their cars but also produced a sociological analysis similar to that of Pierre Bourdieu, in which they highlighted the consumption of cars as an element of social distinction.

The EDF/Renault conflict corresponded to the Touraine/Bourdieu conflict, which meant that resolution of the sociological controversy was linked to that of the technological one:

> The post-industrial society that Touraine believes is coming depends in this particular case not only on the capacity of new protest movements to influence the choices of technocrats but also on the way in which the catalysts in the fuel cells behave ... Although Bourdieu happens to be right and Touraine wrong, this is quite by chance. Although Renault turns out to be right, this is because the heterogeneous associations proposed by EDF engineers disintegrate one by one: the discovery of a cheap catalyst as a substitute for platinum might have proved Bourdieu wrong and rehabilitated Touraine's sociological theory after all. (ibid.: 97)

It should nevertheless be clear that technology is not there to help sociology and provide it with ammunition for its controversies: 'Instead of being someone whose ideas and experiments can be turned to the advantage of the sociologist, the engineer-sociologist becomes the model to which the sociologist turns for inspiration' (ibid.: 99).

For the sociologist who does not take the engineer–sociologist as a model and who thinks that there is a real intellectual specificity in his or her discipline, Callon and Latour's theories can be criticized on three main points: the concepts of the network, the actor and the context.

The Network

The concept of a network, omnipresent in Callon's and Latour's work, is very much a flexible catchall. Like a net that can change to any shape and capture an endless variety of objects, this concept, very fashionable in sociology today, covers widely diverse elements with a single word. For Latour, the network is above all what makes it possible to overcome contradictions.[10] More operationally, it is defined as that which links heterogeneous elements. We are told both that the more extended it is the sounder it is, and that it is only as sound as its weakest link.

This network metaphor is used in four different contexts. First, that of the engineer who builds a network, in which case the two authors question its solidity. Second, the network is an area of circulation (Callon, 1989a: 22). Third, the spatial metaphor evolves, taking us from circulation to the boundary (for example, scientists alter their reasoning when they are inside or outside their networks; they have either a constructivist view of science or an essentialist view (Latour, 1987: 182). The last one is a temporal metaphor ('placing the emphasis on the network means above all suggesting a time frame in scientific work' (Callon, 1989a: 24)).

These different meanings are so diverse that the concept of a network can be used in many circumstances without creating an effect of knowledge. However, in another text Latour clarifies his approach by distinguishing five types of network articulated to one another: instruments, colleagues, allies, the public and, finally, binders (Latour, 1989a: 504). I shall not go into the question of networks of colleagues, allies and the public here. The first, the scientific community, has been apprehended to a large degree by standard sociology of science. Institutional allies have also been studied by the social history of technology. Finally, the public is glaringly absent from the work of Callon and Latour. Even though they categorically refuse the diffusionist model, they have undertaken no work on the uses of science and technology. On the other hand, the question of instruments and, more broadly, of the collection and processing of scientific data is considered at length and indisputably constitutes a strength of these analyses (see Latour, 1985 and 1987, pt 3). Knowledge is essentially accumulation, capitalization. Science has been able to develop only through the proliferation of samples (of rock, insects and so on), measurements, comparisons and the cross-referencing of different data in statistical calculations. There can be no natural science until the different plants on Earth can be brought together in a museum. There can be no geography until people know how to make maps (that is, transmit the same figure to all users) and update them (that is, incorporate new data provided by navigators) and so on.

The fifth network (called 'circle' in the cited text) is defined by Latour as 'binders or links, to avoid the words content and concept whose past is too heavy' (Latour, 1989a: 504). Achieving this circle

> means touching on something harder ... Simultaneously having all the resources mobilized in the other four circles is no joke. Now that all these threads are still scattered, they need to be soundly tied, so that they don't give way to centrifugal forces ... The more there are of them thus assembled, the more necessary it is to find the notion, the argument, the theory that can bind them together. The hardness of the link is what will make the attachment last. (Ibid.: 510–11)

This text can be interpreted in two ways. Either the core of scientific and technological activity is a link, an attachment, and once again we find the associationist metaphor dear to Latour, the explanatory power of which is very weak. Or, while denying the content–context, science–society separation, we nevertheless reintroduce it and have an approach to science and technology that can be very fruitful. We thus avoid both a generalized and sterile associationism and a total separation of an autonomous technoscience, influenced only by a context. Scientific and technological activity clearly has a specificity, but it develops in constant interaction with measurement tools, dialogue with peers, the support of institutions, and user participation. Thus, there are multiple networks that combine in scientific and technological activity.

The Actor

Reconsider the concept of a network. We have seen that the ideas of construction and of an area of circulation are central. We therefore need to ask: who constructs the network and who circulates in it? Answer: the actors. Like the concept of a network, that of the actors is free of ambiguity in Callon's and Latour's work. First, two types of actor can be distinguished: those that are represented, and the spokespersons. Despite the obvious dissymmetry between them, the represented actors have a considerable degree of autonomy; at any point they have the option of following or betraying. These actors may be humans or non-humans.[11] To be able to associate humans and non-humans in the same concept, Latour uses the semiotic concept of 'actant'. For semioticians, 'an actant may embody itself in a particular character or it may reside in the function of more than one character in respect of their common role in the story's underlying oppositional structure' (Hawkes, 1977: 89). In sociology, we can consider that actants comprise all sorts of autonomous figures which are interconnected with others. The other difference between actant and actor is the role of the latter in action: 'An actor is an actant endowed with a character' (Akrich and Latour, 1992: 259).

There is also a second level of actors, the organizers of the network. Yet Callon and Latour have constructed no unified theory of the network organizer as an actor. In some cases there is not even a network organization. Latour, for instance, contrasts two models of innovation: the linear model in which the inventor gradually realizes his or her initial idea, and the whirlwind model, in which that initial idea is of no relevance, has no autonomous force and is not specially propelled by an inventor (Latour, 1992b: 103–4; see also Akrich et al., 1988). The project then develops only if it is adopted by a particular group that adjusts and alters it. Hence, there is no actor–organizer but a sequence of chance events.

In the same book, a few pages down, Latour presents actors engaged in an automatic metro project who are unaware of what they are doing: 'It is not only the sociologist who is unaware of what Matra wants, but also the government minister, the minister's secretary, Lagardère himself, the CEO of Matra. They would also like to stabilize a certain interpretation of what they are and what they want' (ibid.: 149).

Finally, in Callon and Latour we find a third conception of the actor: the strategic actor. 'What is an actor?', they ask. Their answer: 'any element which bends space around itself, make other elements dependent upon itself, and translates their will into a language of its own. An actor makes changes in the set of elements and concepts habitually used to describe the social and the natural worlds' (Callon and Latour, 1981: 286). This strategic actor, not clearly defined as a human, is situated in a world of warriors. Reference is frequently made to Niccolò Machiavelli. The strategic actor forms alliances, undoes those of its enemies, interests any partners who may present themselves, seduces a colleague or a scallop (Callon, 1986).

In this perspective a controversy becomes a power struggle. When a scientist wants to challenge a colleague's statement, he or she has to verify everything that colleague has written, including his or her references, and redo the experiments, for which it may be necessary to create a counterlaboratory: 'So the dissenters do not simply have to get a laboratory; they have to get a *better* laboratory. This makes the price still higher and the conditions to be met still more unusual' (Latour, 1987: 79; added emphasis). Thus, in almost all cases, sceptics give up. In other words, the validity of a statement is recognized only because there are no forces opposing it. In the final analysis, it is the strongest who wins.

Latour clearly illustrates this strategic vision of research by means of the Guillemin laboratory's work on a thyrotrophin-releasing factor (TRF). During competition between laboratories to isolate TRF, Guillemin succeeded in altering the context of validation of results by setting higher criteria of acceptability and having them approved by the scientific community: 'By increasing the material and intellectual requirements, the number of competitors was reduced' (Latour and Woolgar, 1979: 123).

This notion of a strategic actor also appears very clearly in the work of another researcher at the CSI during that period, Vincent Mangematin, who studied two competing technological innovations. He considered that it was the effectiveness of actors' strategies that distinguished them: 'Like in the best detective stories the manoeuvres vary: pre-announcement effects and manipulation of expectations, lobbying public authorities and transformation of technical objects, marketing and setting prices' (Mangematin, 1993: 10). Technical success thus becomes a matter of tactics.

To resolve the contradiction between two conceptions of the evolution of networks – one in which they change in a purely random fashion and the other in which they are manipulated by a strategic actor – Callon proposed the concept of an actor network: 'An actor network is simultaneously an actor whose activity is networking heterogeneous elements and a network that is able to redefine and transform what it is made of' (Callon, 1989b: 93). But the empirical translation of this concept seems to show that the links between the two elements are flimsy. In a case study, Callon (1989b: 190 ff.) used the actor network concept as an equivalent to a strategic actor, while the various competing actor networks were referred to by the name of the researcher who structured them. The concept nevertheless became the standard bearer of this school, more and more frequently referred to as the Actor Network Theory (ANT). The core idea was simultaneously to study actors and the relationships between them. Latour's aim was to start off from his experience of sociology of science, to construct a new sociological theory (Latour, 2005). The French title of his book, *Refaire la sociologie* (literally, 'remaking sociology') is particularly explicit in this respect.[12]

The Context

Innovators' strategic action remains one of the key elements in the explanation of their innovation's success or failure. As regards the electric vehicle studied in the mid-1970s, Callon wrote: 'If failure there was, it was because the forces of EDF were too weak, as they were for the creators of the Concord who, in order to succeed, would have had to control not only steel, exhaust fumes and turbulences, but also the price of kerosene, the movements of citizens living near airports and the democratization of air transport' (Callon, 1981: 397). This statement is confusing. Can innovation be analysed only in terms of power struggles? Are there not factors which innovators can control and others over which they have no control at all?

Despite very clear assertions, we do sense that Callon and Latour are hesitant in this respect. More than once they consider that the concept of a social context needs to be excluded because of the difficulty of distinguishing that which is internal to a technique from that which is external to it (Callon, 1980; Callon, 1987: 100).

Yet three types of relationship between a technological project and its context appear. In the first, 'a technological project is not in a context, it gives itself a context' (Latour, 1992b: 115). In this case the sociologist has to study the simultaneous genesis of a project and its environment: 'If the machine functions perfectly', wrote Madeleine Akrich, 'it makes its definition of the environment "realistic". By contrast, any dysfunction can be interpreted as the intervention of an unexpected factor' (Akrich, 1989: 41). For an innovation to succeed, it gradually has to create an adequate environment.

In Latour's work we also find references to the opposite situation: 'Either Aramis [the automatic metro project] changes to hold its environment and is more likely to exist, or it loses its grip on its environment ... and loses some of its existence' (Latour, 1992b: 173). The situation has therefore switched around: it is no longer strategic actors who adjust their environment to their project, but innovators who adjust their objects to take the context into account. Yet can it be said that the separation between object and context is pointless?

This is where the third relationship comes into play. Our authors explain the failure of carefully prepared innovations (for example, continuous pouring of steel or natural Porvair leather) in terms of an external factor: an unfavourable global economic trend: 'Any innovation implies a favourable environment' (Akrich et al., 1988: 10). Thus, there are elements of context that the innovator cannot control or adjust to. Even if innovators partially define their environment, it still remains an external factor.

In fact, I think that we have to consider that innovators face a dilemma. They act and are acted on. In their field of action (laboratory, factory, marketing campaigns) they develop a technique and impact on the market. When they leave their frame of action they have to make do with what there is, use cunning, be where they are not expected, rapidly abandon a solution in favour of another. They constantly have to move from one level to another, for if they remain at the same level an unexpected change in the market or technology could compromise their project. If, on the other hand, they constantly adjust to new states of the market or technology, they will never produce a major innovation. Innovators constantly have to master their field of action and to question it. Like navigators, they have to master their sails and helm fully and adjust to changes in the winds and currents.[13]

The Limits of Networks

At the end of this long presentation of the French sociology of translation and technology, I think it would be useful to recap. Unquestionably, the main contribution of Callon's and Latour's research is the fact that they succeeded in opening the black box of technoscience; not, like the social constructivism school, to analyse social conflicts within it, but to spread the pieces out on the table with no forethought, without grading the different elements and with the intention of understanding the forces drawing them together into a common network.

The second strength of this theory is that it concentrates the analysis on a few particular moments of technological development (controversy or innovative project), and analyses projects that have succeeded as well as those that have failed. This perspective corresponds to that of social constructivism and cultural technology studies which examine changes due to colonization or technology transfers. These situations have to be at the heart of the study of innovation.

Periods of controversy generally end with a phase of stability. Scientific theory or the technical object then becomes a black box that is used as it is, without any inquiry into its composition. On this aspect of the stability and generalization of artifacts, Callon's and Latour's theses seems debatable.[14] Can the question of rationality be reduced to an actor's strategic ability to impose his, her or its choice? Even if this research has made it possible to challenge many classical studies in the sociology of technoscience based on the notion of pure knowledge or the technical object *per se*, the answer provided in terms of a power struggle hardly seems satisfactory and is unquestionably one of the weaknesses of the theory of technoscience networks. It is therefore hardly surprising that the problematic of Callon and Latour was taken up by sociologists close to Michel Crozier (See Amblard et al., 1996) who saw in it the model of strategic action.[15]

Another point on which Callon's and Latour's research can be criticized is the fact that they disregard the issue of the actor's intentionality, and opt rather for a tactical capacity to grasp opportunities, to 'tighten the screws' of the network. Their choice, to focus on the study of controversies, transformed their view of scientific and technological work, in a sense, by reducing it to an activity of confrontation and conviction.

Can we describe scientific or technological action without the slightest reference to intentionality, to scientists' and engineers' projects? Even if, in the account of a particular discovery, we sometimes find references to initial intentions,[16] the question of intentionality has no place in the model. By contrast, in his study on the Twingo, Christophe Midler distinguishes himself from Callon and Latour's model, even though he uses their notion of *intéressement* and their whirlwind model. In the projects that he analyses there is constant tension between the wish to affirm the identity of the project, and openness to negotiation and compromise with partners both within the firm and outside (Midler, 1993: 66–8). Sociologists of the organizations cited above likewise reinterpret the sociology of innovation model. They stress the fact that at the origin of any change we find a 'temporary and minimal project ... which originally can reside only in the intention to answer a general question, yet which embraces the interests of each of the entities' (Amblard et al., 1996: 156–7). This initial problematization is the work of a translator, a project initiator. I consider that, on the contrary, there is always tension between project and opportunity, and that we should make an effort to incorporate it into the analysis.

The French sociology of translation also refuses the traditional distinction between technological design and diffusion. Yet it believes that once the artifact has been transformed into a black box, diffusion stands to reason, and is therefore of little interest. Although Callon and Latour do sometimes refer to users, in reality they allow them little room in their analyses.[17] As for the social constructivists, they give the same place to manufacturers and users who confront one another within the same antagonistic social groups. I have shown above that even if some of these groups do have a real identity, others are simply virtual. Moreover, although designers and users are involved in the production and utilization phases, their intervention is not the same in the two cases. We have to agree that when users are introduced into the study of technology, the generalized associationist hypothesis underlying the network of Callon and Latour network no longer seems feasible. To be sure, designers and users do cooperate to make a technical object live, but from different positions.

In this review we also need to consider method. To study science (or technology) in-the-making, as the French and British schools wish to, we need to construct an anthropological method. For example, at the beginning of his book on laboratory life, Latour recounted the trials and tribulations of a young ethnographer on his arrival from Africa, in the Guillemin high-tech lab in California. But his ethnographic perspective subsequently disappeared to some extent. His interest in controversies and scientific literature caused him to work far more on written documents and interviews. In *Laboratory Life*, the observation of daily scientific activity was only a part of the work. In particular, Latour presented a remarkable analysis of informal laboratory conversations, scientists' hesitations when faced with unexpected phenomena, the role of chance in finding solutions, and so on. Apart from that he used the usual apparatus of any historian or sociologist doing monographic studies: systematic analysis of written documents (articles, minutes of meetings and so on) and interviews. Latour and Callon subsequently analysed historical controversies (for example, the Pasteur–Pouchet conflict; (see Latour 1989b) or proposed contemporary monographs (the electric vehicle, the automatic metro Aramis, scallop breeding). Sometimes, to distinguish himself better from epistemologists of science who treated only grand subjects, Latour, in his iconoclastic zeal, readily analysed the most futile subjects such as how to weight hotel room keys so that customers remember to leave them behind (Latour et al., 1992) or how Gaston Lagaffe invented the cat-flap (Latour, 1996: 15–24). Although we can but be delighted that this poor tinker from the comic *Spirou* can have a place in the pantheon of the history of science, thanks to Latour, such case studies are purely formal illustrations of a general theory. Quéré notes:

> [Latour] is subjected to the obligation to make observable, as entities of the real world, invisible attributes (such as strategies, interests or operations) given to actors and actions by theoretical formulation ... The peculiarity of this type of analysis of phenomena is that it closes the theoretical formulation in on itself: the theoretical 'construct' requires empirical indicators to support its reality, but far from being drawn from reality, these indicators are entirely made up in relation to the theoretical 'construct' itself. (Quéré, 1989: 112)

Let us revert to the ethnological approach.[18] Latour noted that 'a major difficulty for the observer is that he usually arrives on the scene too late: he can only record the retrospective anecdotes of how this or that scientist had an idea' (Latour and Woolgar, 1979: 172). The researcher then has to reconstruct the events, to build an account. The following two questions arise. What is the observer's role? What creates the coherence of the account? The observer, who is in the situation of a historian and can therefore be referred to as such, has to delimit his or her field of analysis. Latour confined himself to the choice of the actors and selected the same frame as they had. He forgot that the historian's activity is not transparent to the actors' choice. Whether historians like it or not, they make their own choices. Take one of Latour's case studies, long-distance telephony. He developed the thesis that in the late nineteenth century, in order to solve the problem of long-distance transmission, ATT created research laboratories that recruited young physics graduates. These physicists discovered that the triode (electronic amplifier) could become an excellent repeater. Latour drew on historical work on the birth of research at ATT (Hoddeson, 1981). By contrast, other historians who studied the birth of the triode showed that Lee de Forest, who was the inventor of this electronic vacuum tube valve, used it for radio and then suggested its use for telephony to ATT (Aitken, 1985). We thus have two contrasting sequences of translation: fundamental physics → repeater in one case, and radio → telephone in another. It seems strange that it did not occur to Latour, who granted so much importance to scientific controversies, that there could be several histories of the same event, depending on the sources he used. Yet, as Maurice Mendelbaum notes, 'the historian's account is structured by relationships that are clearly indicated in the materials with which he deals' (Mendelbaum, 1977: 117). Hence, the French sociology of translation has not applied to itself the methodological principles that it defined for studying the sciences.

The assessment of the French school is thus both positive and negative: on the one hand, unquestionable contributions; on the other, profound shortcomings that make this theory seem, to me, unsuited to the study of new technologies and their uses. The impossibility of undertaking real ethnographic observation of scientific or technological work shows that we need to draw on other approaches to construct a model for analysing the technique.

THE ROLE OF USERS

Ruth Schwartz Cowan, using the same network concept as Callon, proposed a reorientation of the sociology of techniques by focusing on the consumer. She endeavoured not only 'to place the consumer in the center of the network but also to view the network from the consumer's point of view'.[19] More precisely, she wanted 'to focus on the consumption junction, the place and the time at which the consumer makes choices between competing technologies' (Cowan, 1987: 262–3). Cowan thus drew closer to classical anthropology of techniques, and to the SCOT current which studies both the design of a technique and its use and diffusion. She distinguished herself from sociologists of the Actor Network Theory who, as we have seen, paid little attention to users – although some researchers in this current did undertake some studies on the subject, from an essentially semiological point of view. By contrast, the study of uses has been an important element of media studies and especially the sociology of new media. The focus of these studies has included technical objects. In France this current is often called the 'sociology of uses' (Jouët, 2000). Finally, in organization science the question of users has been at the centre of certain studies on the computerization of firms. A topic that has mobilized the social sciences to a large degree is complex

Semiotic Approaches

In a metaphoric approach – the machine as text – Woolgar compared the user to a reader. During the design of a machine, engineers set parameters for the user's actions. So 'the machine effectively attempts to *configure* the user' (Woolgar, 1991: 61; original emphasis). To design and produce a new object requires defining 'the identity of putative users, and setting constraints upon their likely future actions' (ibid.: 59). The final machine encourages only certain forms of access and use. Users are considered as having no power compared to designers; they are outsiders. Hence, in Woolgar's approach, users appear only through the representations that designers have of them. By studying how technological objects constrain the ways in which users relate to things, there is always a risk of tilting over into technological determinism. In fact it is strange that Woolgar, often presented as the initiator of a semiological approach, seems to have overlooked Stuart Hall's seminal article entitled 'Encoding, decoding in television discourse' (1981). In this work Hall showed how, in the case of television, the viewer can decode a programme in a different mode to the one chosen by the producer for encoding it. The TV viewer may either follow only some of the message or else reject it entirely. This autonomy was to be the base of many studies on the reception of television. Sociologists of work also observed for a long time the wide gap between work as prescribed by machines and organizations and the work actually carried out by employees.

Some authors, like Hugh Mackay et al., criticize Woolgar's approach, and especially the implicit determinism in the concept of configuring the user. They complete his approach by studying the equally important 'decoding', that is, the work done by users (readers) to interpret the machine (text). This symmetric approach led them to believe that whereas designers configure users, 'designers, too, are configured by their organizations and by users' (Mackay et al., 2000: 752).

The same semiotic model is found in Akrich's work. She argues that innovators have a representation of users' tastes and projects, and that from there they 'embed ... this vision of the world in the technical content of the new project'. She calls this work a 'script' or 'scenario': 'Like a film script, technical objects define a framework of action together with the actors and the space in which they are supposed to act' (Akrich, 1992: 208). As a result, the script contained in technologies attributes specific actions to users. In this definition of the script, a key element is the definition of the skills and behaviours of future users. As users are heterogeneous, designers choose an easy option: they consider their own preferences as representative of those of users. In another article, Akrich refers to her perspective as an 'I-methodology... whereby the designer replaces his professional hat by that of the layman' (Akrich, 1995: 173).

At first the notion of a script resembles the idea of configuring the user, dear to Woolgar. But Akrich refuses to favour the designer's point of view. She believes that one should study not only the user's project but also the real user, and contrasts the world embedded in the object by the imagined user, with the one embedded in it by the real user. She then sets out to articulate design and uses, sociology of innovation and sociology of action (Akrich, 1993: 55). From the point of view of uses, her main focus has been the question of users' cooperation with the device. The examples that she studies primarily concern interfaces with the technical object. In the case study to which she often refers, of connectors with the cable TV network, she examines the different affordances of the connector, the functions proposed, and human–machine dialogues. Very little attention is paid to the new TV practices allowed for by cable TV, such as the new programmes received and the individualization of reception, of particular interest to specialists of the reception of television. Thus, Akrich's analysis of uses remains contingent on technical objects. She tries more to answer the question of how they are used rather than why they are used.

Sociology of Media and New Media

Whereas in the contemporary sociology of techniques the question of uses has been secondary, in media studies it has been central. The 'uses and gratification' approach was a turning-point since it overturned the classical paradigm of the effects of the media. The new research programme consisted in analysing what 'individuals do to the media' and not what 'the media do to individuals'. Elihu Katz and his colleagues argued that 'people bend the media to their needs more readily than the media overpower them' (Katz et al., 1973: 164–5). This approach of gratifications studies was to evolve as it focused on sense-making activities of audiences and encountered the cultural studies current of which Hall was one of the most eminent researchers. As David Morley notes, by studying 'local processes of situated consumption', researchers discovered that 'local meanings are so often made within and against the symbolic resources provided by global media networks' (Morley, 1993: 17). Hence, media studies evolved from text to context, from semiotic analysis to social analysis. The period of text-reader reception was replaced by ethnographic studies of everyday life. With the concept of media articulation, some researchers such as Roger Silverstone contrasted media-as-text and media-as-object, symbolic messages on the one hand, with technological objects located in particular spatio-temporal settings, on the other: 'In short, people are always both *interpreters* of the media-as-text and *users* of the media-as-object, and the activities associated with these symbolic and material uses of media are mutually defining' (Livingstone, 2003: 14; original emphasis). While active audience theory became the mainstream of media studies, interactions between designers and users became the new approach of ICT studies. This was the context in which British researchers developed their work on ICTs and everyday life. Silverstone and Hirsch focused their work on household and family life. They analysed households as a 'moral economy', a unit actively engaged in the consumption of objects and meanings, and saw ICTs as 'crucial to the household's more or less successful achievement of its own identity, integrity and security ... and for the household's ongoing engagement with the commodities and symbols of the public sphere' (Silverstone and Hirsch, 1994: 6). They showed how this domestication of ICTs in everyday life is organized in four phases: 'appropriation', 'objectification', 'incorporation' and 'conversion'. With appropriation, the technology leaves the world of commodities, as individuals or households become its owners. This phase is close to the 'consumption junction' studied by Cowan. With objectification, technologies find a physical disposition in the familiar environment and enter into a process of display. A spatial differentiation appears 'private, shared, contested; adult, child, male, female, etc.' (ibid.: 23). During the third phase, technological objects are used and incorporated into the routines of daily life. This process is accompanied by 'constant work of differentiation and identification' within the household. Conversion is used to describe the processes in which the use of technological objects 'defines the relationship between the household and the outside world' (ibid.: 25). Users show themselves to others by using the technique; they talk about it to them.

This concept of domestication is not only the transfer of the approach of an active consumer to the world of objects. The starting-point is no longer encoding or the technical object's design, nor even user–machine interactions, but the dynamics of users' world. The user is in the centre of social and cultural relations. Silverstone's approach resembles that of Sherry Turkle, whose 'subjective computer' is 'the computer seen in its relationship to *personal meaning*' (Turkle, 1982: 177; original emphasis): 'Working with computers can be a way of "working through" powerful feelings' (ibid.: 174).

The French tradition of sociology of use has slightly different origins from the British tradition (Jouët, 2000), and it took a long time for media studies to get off the ground in France. On the other hand, France has a longer tradition of studies on the uses of ICTs. These started in the 1980s, with the appearance of the videotext.[20] Sociologists who observed the first uses of the Minitel pointed out the wide gap between the expected and the actual uses of this device. Promoters of the new technology had imagined a rational user who would consult databases, whereas in fact direct dialogue (message) services accounted for most of its use (Jouët et al., 1991). This French sociology of uses studied most of the ICTs, especially in the family context (Jouët and Pasquier, 1999). Three main strands of research can be identified: first, the genealogy of uses (Perriault, 1989; Flichy, 1995); second, take-up or appropriation of the new technique by an active user, which has both a cognitive dimension (De Fornel, 1996) and an identity dimension (as Jouët (2000: 502) noted, 'appropriation is a process, it is the action of creating one's self'); and, finally, the new communities of users that formed around ICTs (Hert, 1997; Beaudouin and Velkovska, 1999).

Along with the two British and French sociology of communication schools, more recent cultural studies work has also focused on technical objects. Paul du Gay, Stuart Hall and their colleagues examined cultural artifacts, such as the Walkman. Their aim was a comprehensive study of a technical object, including both its production and its consumption. In fact they adopted the point of view of consumption, and considered production essentially via marketing and advertising. But of most interest is the fact that they actually noticed the cultural aspect of artifacts. This corresponds to the perspectives of cultural technology mentioned at the beginning of this chapter, which are applied to contemporary mass consumer products. Du Gay, Hall and their colleagues define culture as 'a shared framework or "map" of meaning which we use to place and understand things to make sense of the world' (du Gay et al., 1997: 8). The adjective cultural associated with an artifact relates to a distinct set of social practices and to certain kinds of people, places and situations. But this map of meaning is associated not only with real practices but also with the *mise en scène* of these technologies in literary or cinema fictions and, more broadly, in the media and advertisements. Other sociologists also tried to include real or represented users as well as non-users in their reflection. Sally Wyatt identifies four different types of non-user: the 'resisters', people who have never used the technology, because they do not want to; the 'rejecters', people who have stopped using the technology, because they find it boring or expensive; the 'excluded' who cannot get access for social or technical reasons; and the 'expelled' who no longer use the technology because of cost or institutional access (Wyatt, 2003: 76). We could add resisting consumers, intermediaries between users and non-users, who have 'resisted, modified, and selectively adopted technologies' (Kline, 2003: 51).

These different approaches have the advantage of highlighting the fact that, in a particular society, the meaning given to a technical object and the uses envisaged constitute a shared framework. The relationship to the technical object is profoundly collective. It associates users, non-users, artists and advertisers, among others.

Information Technologies and Organization

Whereas the semiotic approach examined users through the interface with the machine, British media studies and the French sociology of uses have shown little interest in the direct relationship with the object. Instead, they have focused on processes of appropriation and sociality related to the technique. The last research current that I wish to present studied the articulation between technologists and users in a different context: computer use in business environments.

The community of management information systems has been studying the relationship between computer technology and organizations for a long time. Wanda Orlikowski has shown that we cannot accept a model that considers technology as an external force with deterministic impacts on organizational properties, nor as one that emphasizes individuals' ability to make choices to interact with technology. The duality of technology has to be integrated into the analysis (Orlikowski, 1992). In other studies she uses the notion of a technological frame proposed by Pinch and Bijker in a socio-cognitive perspective. It is tacit knowledge that facilitates understanding and avoids ambiguity. This is not purely cognitive, it is also practical: 'Technological frames are the understanding that members of a social group come to have of particular artifacts, and they include not only knowledge about the particular technology but also local understanding of specific uses in a given setting' (Orlikowski and Gash, 1994: 178). Effects of congruence (alignment) or of incongruence can be observed between technological frames. These authors show, with regard to their observation of the introduction of Lotus Notes, that frames of 'technologists' and 'users' differ. They distinguish between the 'nature of technology', which refers to images of the technology, its capabilities and functionality, 'technology strategy' (view of the organization which acquired and implemented the technology) and 'technology in use'. Technologists and users do not have the same approaches. Technological frames are incongruent. To avoid this break, Orlikowski and Gash identify a set of activities called 'technology-use mediation': deliberate and organizationally sanctioned intervention, 'to adapt a new communication technology to that context, modifies the context as appropriate' (Orlikowski et al., 1995: 424).

In another article, Orlikowski distinguishes 'technological artifact' and 'technology-in-practice'. The former includes material and cultural properties and common activities, while the latter refers to 'the set of rules and resources that are (re)constituted in people's recurrent engagement with technologies at hand' (Orlikowski, 2000: 407).

The thesis of Orlikowski and her colleagues theses are interesting in so far as they grant far more room to the technical object and technologists than does the sociology of communication. Unlike the sociology of communication, they open the black box, albeit imperfectly. In the encounter between technologists and users, Orlikowski's technologists are essentially the prescribers and installers of the technology, not the designers. Their role is to institute technological and organizational change. Use is analysed in situations – of work in this case – as it was in family life situations by Silverstone or Jouët and Pasquier.

Diverse User Figures

To conclude, the research presented in this chapter clearly illustrates the complexity of technologies-in-practice. Uses are not deduced from encoding by engineers or an initial script. The notion of use is moreover many-sided. It relates to the interface with the machine, to social and technical representations, to the position in the space of daily life, to rules of usage, to the social practices in which the technology is embedded, and to the resources mobilized by users.[21] When they seek to combine the study of use with that of technical design, these studies often have an oversimplified view of the process of designing a technical artifact. Those whom they call designers, technologists and producers represent only some of the actors involved in design work. Likewise, there are numerous user figures: prescribers, buyers, professional users, domestic users and so on. The approach to technology and its uses that I propose is intended to incorporate all the dimensions of design work and uses, and to give a place to the multiple actors of technical action.

NOTES

1. This is the title of the two articles on technology in the encyclopaedia of ethnology published in 'La Pléiade' (Poirier, 1968).
2. See Pacey (1983) and in the Science, Technology and Society (STS) tradition Akrich (1992: 205–24).
3. For a presentation of the new social approach to science, see Pestre (1992).
4. This model was refined a few years later by Bijker (1995).
5. However, more recently Pinch has argued that it is 'more useful to see closure as something that is continually in operation' (Pinch, 2001: 398).
6. Bijker studied a case of this kind in an article on fluorescent lights. He showed how this artifact was the result of a negotiation between manufacturers and the public lighting authorities. See Bijker, 1992: 75–102.

7. On the debate around the SCOT, see the article in *Technology and Culture* by Nick Clayton (2002) and Bijker and Pinch's response (Bijker and Pinch, 2002).
8. Michel Serres noted: 'Disorder is in the order of things and order in their exception' and, a few lines down with regard to science: 'Its limit and its boundary is this real, filled with disorder, where its language is scattered and dissolves into sounds, its ground is the island of the informed real, an improbable domain, a pinpoint where the logos is sown' (Serres, 1980: 158–9).
9. 'By translation, we understand all the negotiations, intrigues, calculations, acts of persuasion and violence, thanks to which an actor or force takes, or causes to be conferred on itself, authority to speak or act on behalf of another actor or force', explained Callon and Latour, 1981: 279.
10. 'If technoscience may be described as being so powerful and yet so small, so concentrated and so dilute, it means it has the characteristics of a *network*' (Latour, 1987: 180; added emphasis).
11. For a critique of the point of view in which non-humans are actors, see Collins and Yearly, 1991.
12. I am unable, within the scope of this book, to present this theory in full, along with its relation to the sociology of Gabriel Tarde. I have limited myself to the work of Callon and Latour in the sociology of science and techniques.
13. On this subject of the innovator's intervention, see Flichy (1991).
14. This aspect of Callon's and Latour's theses has generated many controversies. See: Isambert (1985: 485–508); Thuillier (1987: 506–11); McGuire and Melia (1989: 87–99); Boudon (1990).
15. In more recent writings, Latour distances himself from this interpretation of his model (Latour, 2005).
16. For instance, Latour writes 'Guillemin's initial decision was to determine the structure' (Latour and Woolgar, 1979: 124).
17. We shall see that later another researcher in this team, Madeleine Akrich, did nevertheless address this question. She noted that 'as soon as the technical object becomes an object of use, it stops being of interest to the analyst who sees the user only as the unproblematical extension of the network constituted by the innovator' (Akrich, 1993: 36).
18. A critical analysis of Latour's and Woolgar's theses can be found in Lynch (1993: 93–102).
19. This theme has been developed by feminist scholars. Oudshoorn and Pinch (2003: 4) thought that Cowan played an important part in this 'turn to the users'.
20. *Réseaux* was the main journal in which this work was published. An English version (*Réseaux: The French Journal of Communication*) was published from 1993 to 1999.
21. Use is also distinguished from purchasing and consumption. Moreover, it is difficult to measure. While purchasing can be measured in terms of level of equipment, and consumption in terms of volume or value, it is more difficult to quantify use, for which the only general unit of measurement is time.

4. Socio-technical action and frame of reference

How can we define an approach to technological innovation that adopts both the designer's and the user's point of view? To construct this new socio-technical model of innovation, I draw upon the sociology of action as well as interactionist sociology. I show that innovators' and users' action fits into a socio-technical framework and may be either strategic or tactical.

If we are to define a new approach to the study of technology and its uses, we need to consider in more depth the activities of the actors involved, the interaction between them and their frames of action. Two research movements, ethnomethodology and interactionist sociology, offer useful perspectives for clarifying such concepts. The first two sections of this chapter will therefore focus on these sociological schools. From there I develop my own model for the analysis of technology and its uses. In particular, I examine the frame of reference of action and the various modalities of strategic and tactical action.

FORMS OF SCIENTIFIC AND TECHNOLOGICAL ACTION

It seems appropriate to start by exploring the internal dynamics of scientific and technological action, that is to say the interpretations, deliberations and interactions that human actors develop with regard to themselves or to others. Since ethnomethodology[1] considers that there is no fundamental difference between scientific action and ordinary action, the characteristics of the latter need to be considered first (see Pharo and Quéré, 1990).

At the starting-point of ethnomethodological thought we find questions relating to the procedures used by individuals to give meaning to their own and others' actions. All the objects of the social world, says Alfred Schutz, are constituted within a framework of 'familiarity and pre-acquaintanceship supplied by a socially constituted "stock of knowledge" at hand (Schütz, 1962: 7, cited in Heritage, 1987: 230). But this typified knowledge is open to constant revision. In short, ethnomethodology wants to answer Harold

Garfinkel's question of 'how men, isolated yet simultaneously in an odd communion, go about the business of constructing, testing, maintaining, altering, validating, questioning and defining an order together' (Garfinkel, 1952: 114, cited in Heritage, 1987: 232).

Garfinkel's answer can be summarized, according to John C. Heritage (Heritage, 1987: 247–8), in three points:

1. the situation of action is not united in a standardized and determining context of activities; on the contrary, it is capable of constant transformation. Action and context are determined and elaborated mutually;
2. the norms whereby situations and actions can be recognized are not 'rigid templates'. They should be understood 'as elastic and revisable resources which are adjusted and altered over the course of their application to concrete contexts'; and
3. these norms are not the driving force behind behaviour. Ethnomethodology is more interested in normative conventions which are essential resources for interpreting frames of action.

By applying this type of approach to science, Garfinkel introduced a fundamental break in sociology. The aim was no longer to examine how social structures impacted on scientific work, or how scientists managed to convince their colleagues of the validity of their theories, but rather to study scientific work itself.

Sometimes, as in the discovery of the optical pulsar, ethnomethodology may gather the results of observations which make it possible to analyse key moments in scientific work. In this particular case, Garfinkel, Michael Lynch and Eric Livingston used recordings of conversations between researchers on the night they made their discovery, as well as the handwritten notes of these researchers. They thus had information on ordinary scientific activity: gestures, settings of laboratory equipment, measurements which were conclusive and those which were not, results which were used to construct the theory and those which were considered as insignificant. The actors' doubts, uncertainties and excitement were also visible. Yet not all these elements appeared in the scientific publication that reported the discovery. As Garfinkel et al. noted, the 'pulsar is depicted as the cause of everything that is seen and said about it. It is depicted as existing prior to and independently of any method for detecting it ... the observers' practices were "naturalized"' (Garfinkel et al., 1981: 131–8). Thus 'science as it is done' is effectively different from 'science as it is represented', from that which appears after the discovery. The ethnomethodological approach seems particularly fertile here in that it allows a description of the research work not usually revealed in scientific publications.

In most cases, however, ethnomethodology does not have access to such a wealth of material for studying scientific activity. It generally has little more than a few sequences of ordinary scientific work. Some researchers have observed daily work in a laboratory and have made audio or video recordings of all the operations carried out by scientists in their work. They have, in particular, sought to analyse the discrepancy between ordinary scientific practice and the principles of method which are meant to govern that practice. Their objective has not been to work on a scientific innovation as a whole, but to focus their investigation on micro corpuses or possibly to demonstrate a number of 'destabilizing experiences'. Garfinkel, for example, asked one of his students to handle test tubes for a young chemist who was paralysed, seated beside him. Through observation Garfinkel was thus able to determine the place of the body in laboratory work. Handling objects is an element in a chemist's reasoning, just as playing an instrument is at the heart of musical activity.[2]

Even if science in-the-making does not correspond to the description given by the scientific article reporting it, does this allow us to say that it is merely the result of pure *bricolage*, of a more or less unexpected encounter between different phenomena which are then organized into a network. Or, by contrast, does the scientist have an intention, a project or an action plan?

In order to answer this question, we shall first examine the way in which ordinary action is described. Paul Ricoeur considers that, in contrast to a mere physical movement, action can be described by the following elements: it implies an aim and relates to motives; it is accomplished by agents in specific circumstances: 'Moreover, acting is always acting "with" others. Interaction may be in the form of cooperation, competition or opposition. The contingencies of interaction are then similar to those of circumstances in so far as they are by nature helpful or adverse' (Ricoeur, 1991: 110; see also Ricoeur, 1977).

Having established the characteristics of human actions, it seems necessary to examine in more detail the relations which exist between action plans and action underway. Lucy Suchman, who worked on the question with respect to cognitive science, developed an interesting analysis on this point (Suchman, 1987). She considers that, unlike some supporters of cognitive science, we cannot identify the action plan of a computer program. We cannot consider that a project describes each of the actions that will constitute a sequence. Here again, ethnomethodology proposes the opposite: 'Plans do not describe the mechanism whereby the action is produced ... rather, they are the constructs of common sense, produced and used by the actors engaged in their daily practices' (Suchman, 1990: 157). To clarify the idea that 'plans are not the determinants of action' (ibid.: 159), but rather resources that actors mobilize, she takes the example of people crossing rapids in a canoe. Before confronting the river they will probably draw up a navigation plan, but in the heat of the action the plan will be forgotten: 'The aim of the plan is not to enable you to cross the rapids, but rather to prepare you so that you are in the best conditions possible to use the know-how on which your success ultimately depends' (ibid.: 158).

The example of the canoe and the rapids may seem somewhat simplistic when the problem concerns not coordination between two individuals and the strength of the current, but rather coordination between a large number of persons cooperating in big organizations with a strong technological culture. It is therefore useful to consider various studies of information systems in large firms, which reach similar conclusions. Elihu Gerson and Susan Star studied the sequence of data processing in a health insurance firm. They noted that problems which arose on a daily basis were dealt with by means of knowledge which was 'tacit, not codified and often not codifiable' (Gerson and Star, 1986: 265). Hence, it is impossible to anticipate all the configurations which may arise or to define a rule applicable to each case. In real-world situations, employees who must make decisions concerning rates for refunding specific illnesses or treatments have to articulate a variety of information, either alone or in co-operation with one another.

Information systems have a dual nature. On the one hand they are open systems, for it is impossible to describe them in full or to plan all possible configurations. On the other hand, in order to function the system has to be partially and temporarily closed: 'The problem ... for information systems, is to develop ways of designing systems that continuously evaluate their own conditions, and restructure themselves as changing circumstances require different patterns of local closure' (ibid.: 267).

Mike Robinson, who has conducted numerous studies on the use of corporate information systems, considers that strategic plans or work programmes should not be seen as being valueless. These 'procedures are more like advice than algorithms' (Robinson, 1993: 189).

FROM NEGOTIATION BETWEEN SOCIAL WORLDS TO BOUNDARY OBJECTS

Such reflection on planning and procedures of scientific and technological action show the extent to which it is necessary to consider action and context together. In Joan Fujimura's words, 'we should study science and society as both constitutive of and consequences of action' (Fujimura, 1991: 222).[3] Structures are in fact the result of past actions which have fossilized but remain present. This approach is typical of interactionist sociology, which we shall now consider in more detail.

Interactionist sociology is interested in the study of processes and collective actions rather than in established facts. As Everett Hughes says, 'society is synonymous with interaction, but interaction with certain persons rather than with others' (quoted by Baszanger, 1992: 56). Processes of coordination of activities do not take place in a random fashion but rather in relation to specific contextual elements.

At the heart of the interactionist approach we find the idea of constant tension between indeterminacy and structure: 'It is precisely this continual necessity for reassessment that permits the innovation and novelty of human life. ... Innovation, in fact, rests upon ambiguous, confused, not wholly defined situations. Out of ambiguity arises challenge and the discovery of new values' (Strauss, 1969: 26). It is this lack of definition which leaves a large degree of latitude to individual action. The latter, in turn, is situated in a macro-social framework which impacts on the conditions of action.

This articulation of the micro and macro social appears in a key concept of the interactionist school: the reference group. Tamotsu Shibutani defines this notion as:

> the matrix through which one perceives his environment. ... Since he defines objects, other people, the world, and himself from the perspective that he shares with others, he can visualize his proposed line of action from this generalized standpoint, anticipate the reactions of others, inhibit undesirable impulses, and thus guide his conduct. (Shibutani, 1955: 564)

Reference groups share perspectives which allow for collective action manifested in the construction of social worlds. These are not static, they evolve constantly. Participation in these worlds is usually fluid and exists to a greater or lesser degree. Howard Becker calls those at the centre of a world 'entrepreneurs' (Becker, 1963). Moreover, individuals may belong to several worlds simultaneously.

The concept of a social world must be clearly distinguished from that of an organization. The boundaries of a social world are fuzzier and, more importantly, social worlds overlap with organizations. An institution may comprise several social worlds, while in other cases a social world may be common to several organizations.

Different social worlds confront one another in the definition of a question – for example, a professional practice – in specific situations: arenas. In these arenas, 'different subjects are debated, negotiated and the representatives of different worlds or sub-worlds confront one another and attempt to manipulate one another' (Strauss, 1978: 124). Strauss uses this concept to study various conflicting therapies in a psychiatric hospital. In this respect he speaks of a 'battlefield' and shows that each of these opposing social worlds proposes a specific organization of the hospital and, more broadly, a structuring of psychiatric activity as a whole (quoted by Clarke, 1991: 129).

This approach, used by certain sociologists of science (Clarke, 1990), seems particularly fertile. It avoids the errors of certain earlier theories which implied that science is constructed either by chance or as an outcome of scientists' manipulatory skills. The theory of social worlds and arenas makes it possible to analyse conflicts and negotiations during the production of scientific and technological artifacts. Moreover, this theory remains open since the sociologist does not use predefined categories. He or she uses research results to determine the social worlds involved in the process. Other analysts may, of course, contest these choices and propose an alternative map of the social worlds.

Furthermore, the interactionist approach makes it possible to avoid the dead-ends of a sociology of institutions which studies medical or scientific institutions, for instance, without attempting to analyse illness or science. In order to do so, other social worlds need to be integrated into research. The sociology of health, for instance, must take the patients into account, just as the sociology of technology must consider the users of technology.

In the interactionist approach, the concept of arenas is essential. They are places of confrontation and cooperation between social worlds. This notion has been adopted and clarified by Susan Star and James Griesemer, who studied the establishment of a zoology museum on the American west coast. At the starting-point of their analysis we find the following question: how can actors be made to cooperate, when new objects have different meanings in different worlds? How can different meanings be reconciled? To answer this question they introduce the notion of boundary objects. These are objects situated at the intersection of several social worlds, which meet the needs of all these worlds simultaneously: 'They are objects which are both plastic enough to adapt to local needs and the constraints of the several parties employing them, yet robust enough to maintain a common identity' (Star and Griesemer, 1989: 393).

In the case studied by Star and Griesemer, cooperation between scientists and amateurs (who were partly responsible for collecting the animals), administrators and patrons, hired hands and so forth, was necessary. To organize this cooperation, a common ideology was developed, that of the conservation of California's natural heritage. Work methods were also established. The scientists wanted animals in a perfect condition, together with a maximum of information on their natural habitat, and the trappers had a difficult and sometimes dangerous job. Procedures for capturing wild animals, which took into account the constraints of both parties, had therefore to be agreed. Similarly, plans for collecting information on the animals' environments had to be devised: 'These methods thus provided a useful lingua franca between amateurs and professionals' (ibid.: 407).

Cooperation was also strengthened by the financial exchange between the museum and the trappers.

The boundary object concept is similar to another one, proposed by Louis Bucciarelli: the object world. This term refers to 'the domain of thought, action and artifact within which participants in engineering design, move and live when working on any specific aspect' (Bucciarelli, 1994: 62). The definition of a technical object gives rise to negotiations between the different object worlds, and the actors often have to reach compromises. In the case of the construction of a photovoltaic system for a solar house, Bucciarelli shows that engineers proposed a voltage of either 12 or 96 volts, depending on their object world, and that they finally decided on the compromise of 48 volts: 'Hence design is best seen as a social process of negotiation and consensus, a consensus somewhat awkwardly expressed in the final product' (ibid. 21).

The observers of project concepts (Clark and Wheelwright, 1992) came to similar conclusions. Successful projects are the result of constant compromises with the different actors who have specific competencies. There is a continuous play of negotiation and interaction. This idea of negotiation, compromise, innovation induced by opportunities and ad hoc agreements is found both in Bucciarelli's work and in that of the specialists of project concepts. The boundary object concept is more subtle. It is not only a compromise but is closer to what French ergonomist Pascal Béguin calls a 'common world'. He defines it as a map with several entries where everyone finds what they need. It is 'a system of different positions, designed to relate debates between people to the complexity of reality rather than to their relationships' (Béguin, 2007: 180).

The schema of a boundary object or of common worlds also differs from that of the French sociology of translation. We no longer have an actor trying to impose his or her own vision of the world on the other actors that he or she needs. On the contrary, we witness the elaboration of a compromise. While the Latour model 'can be seen as a kind of "funneling", reframing or mediating the concerns of several actors into a narrower passage point', that of Star and Griesemer 'is a many-to-many mapping, where several obligatory points of passage are negotiated with several kinds of allies' (Star and Griesemer, 1989: 390).

THE SOCIO-TECHNICAL FRAME OF REFERENCE

The ethnomethodological and interactionist approaches thus provide a number of theoretical elements for founding a new approach to technology and its uses. Such an approach has to meet the following four objectives:

1. Technology and society must be included in a single analysis without one term taking preference over the other, unlike the approach of cultural technology which favours the technical dimension, or that of social constructivism (Pinch and Bijker, 1987) which considers society as the decisive factor. The aim is not only to articulate the two poles, but also to see how numerous social worlds interact (that is, those of engineers, users, industrialists, service suppliers, technicians, merchants and so on).
2. Technology, both in its design and its uses, must be at the heart of the analysis. It is not enough to analyse a single term or to study inventors and consumers independently of the technological activity that they perform. We need to examine the daily activity of laboratories, or the practices of users at home or in the workplace.
3. Research must focus not on technological facts but on technological action, on the intentions, projects and deliberations which precede action, on the course of action itself and primarily on the interaction between different actors as well as between these actors and the technological object.
4. This interaction is possible only if there is a degree of stability in the relations between actors themselves, between actors and the technological object, and in the functioning of the object. Hence, we should be able to account for phenomena of predictability concerning the technological act.

In presenting my analytical model I shall start with this first point and will draw upon Erving Goffman's reflection. Goffman considered that any social event is organized by a primary framework, 'that is seen as rendering what would otherwise be a meaningless aspect of the scene into something that is meaningful' (Goffman, 1974: 21). He distinguished two types of frame, the natural frame and the social frame:

> One, more or less common to all doings, pertains to the patent manipulation of the natural world in accordance with the special constraints that natural occurings impose; the other understanding pertains to the special worlds in which the actor can become involved which, of course, vary considerably. Thus, each play in checkers involves two radically different bases for guidance: one pertains to quite physical matters – to the physical management of the vehicle, not the sign; the other pertains to the very social world of opposing positions that the play has generated. (Ibid.: 23–4)

Although the question of technology is absent from Goffman's work, I believe that we can use this theory of a frame of action and consider that all technological activity is situated in a frame of reference. The actors in a technological operation mobilize a particular frame which enables them to perceive and to understand the phenomena they witness and to organize their own action. In contrast with the notion of a 'technological frame', which is used by the social constructivists and Wanda Orlikowski and is specific to each social group, the notion of a frame of reference which I propose here is common to all the actors, without necessarily being unique. Several frames of reference may coexist or oppose one another, but the actors in a technological operation are always placed in relation to a frame of reference.

To use the concepts of interactionist sociology, we can consider that each social world has its own frame of reference. But as soon as there is a need to plan interaction between actors belonging to different social worlds, an arena is established to create a common frame of reference. This is a boundary frame peculiar to the different actors collaborating in a technological activity (inventors, engineers, technicians as well as users). It may change over time, for while it has to be rigid enough to maintain the coherence between actors, it must also be flexible enough to take into account each one's specific projects.

To clarify this notion of a frame of reference, imagine the technical object as something unfamiliar and strangely distant, then consider a visitor to a technology museum or even a browser in a flea market. He (or she) discovers a machine composed of metal and wood, about which he knows nothing. He then inquires and learns that it is a machine for sending tubular letters. Without compressed air, without a cylinder in which to place the letters, without a network of tubes, this machine cannot function. It is a beautiful copper and wooden object and could be displayed for its aesthetic value. But, removed from its frame of reference, it no longer has any use. If our visitor continued to browse around he would find other simpler objects which could be transferred from one frame of reference to another. He might, for example, find a chandelier in which he could replace the candles by sockets and bulbs before connecting it to an electric circuit. For a more complex object this type of transfer is impossible. We could therefore say that, removed from its frame of reference, a technological object is no more than an archaeological relic. In order to function it needs technological artifacts to which it is bound. It also needs a method of operation and, more generally, know-how as well as other human actors capable of repairing it or of constructing a replacement when the original is finally defunct.

If our visitor to the flea market found a machine which no longer functioned because the required energy source was not available, he would say that it was out of use. If, on the other hand, he found an object that was broken or in a state of disrepair, he would say that it was unusable. Yet, in both cases it would not be the use itself that would pose a problem, but the functioning of the technical artifact. Such a shift in meaning between functioning and use is possible because the two components of the technological artifact are linked. Functioning and use are the two sides of the same coin. The frame of reference can therefore be subdivided into two distinct but articulated frames: the frame of functioning and the frame of use. These two frames have the same relationship as that linking the signifier and the signified in semiology. Together, the two form what I have called the 'socio-technical frame'. If we take the comparison with semiology further, we could say that the socio-technical frame is similar to the sign. There is no necessity in the articulation of a frame of functioning and a frame of use. There is neither determinism nor the reflection of one structure by another. Nevertheless, the relations between functioning and use were constructed over time.

The frame of functioning defines the body of knowledge and know-how mobilized or mobilizable in a technological activity. This frame is not only that of the designers of a technological artifact, but also that of its producers, its repairers and its users. The latter can mobilize the frame when they want to 'open the black box', tinker with it or modify the machine. But for other users this frame of reference simply enables them to orientate their search for advice or help. Any user knows, for example, that the national electricity and telephone networks transport fundamentally different electric currents.

As with any boundary frame, this one also has peculiarities for each of its users. Josiane Jouët describes the two levels of user involvement within a frame of functioning in the case of micro computing: that of the uninitiated and that of professionals:

> For most uninitiated practitioners, familiarity with the technology is limited to the acquisition of know-how and rudimentary notions, to the capacity to operate the tool and to a vague familiarization with the logic of computing. These individuals are, one could say, semi-literate in computing or functionally literate. They do not meet criteria for real computer literacy because their level of knowledge remains too elementary. Yet they are not bewildered when faced with a computer. They use it with ease, even if such use is limited. For uninitiated users, the technology remains a black box which is merely a tool and not an object of knowledge. (Jouët, 1990: 217)

Uninitiated and professional users do nevertheless have the same functional relationship with the machine. In another example, that of the motor car, the frame of functioning, which is that of mechanics, limits the field of possibilities of the design office, factory staff, garage mechanics and users who want to maintain or even modify their vehicle. The frame of reference relates to a variety of knowledge (in this case on the principles of an internal combustion engine) but also to know-how (taking apart and reassembling an engine). Another essential element is the fact that the technological activity is, to a large extent, mediated by various tools and instruments. There is no direct human–machine contact. The individual is faced not with a machine with a capital M, but with a keyboard, gauges and tools which constitute both the frame of functioning of the technology and the frame of its use. From this point of view, the study of human–machine interfaces, one of the activities of ergonomics, is interesting. The development of interfaces is unquestionably one of the aspects of the creation of a frame of functioning. It is even the most visible aspect of this frame. But at the same time, these interfaces must take into account the frame of use. The interface is therefore not the point of articulation between the designer and the user, as Madeleine Akrich says, but rather that of the frame of functioning and of the frame of use. Take, for example, the definition of a new telephone terminal. The ergonomist will participate in the creation of the keyboard – for example, position and size – but must also include, in the technological object, available information on its use, so that the most common functions may be easily accessible.

The frame of use is not limited to users' activities. This is rather a notion which relates to political economy's use value. It describes the kind of social activities proposed by the technology, the integrated routines of daily life, sets of social practices, kinds of people, places and situations connected to the technical artifact. It gives the social meanings of the technology, which naturally has a symbolic dimension, as Sherry Turkle reminds us with her 'subjective computer'. The designer, like the user but also the non-user, thinks about use. It is not because non-users do not use the technique that they are not a stakeholder in the frame of use. This frame shows them what the purpose of the technique is, and those indications enable them to decide whether to use it or not. As for designers, they are interested not only in the performance of the new device they are designing, but also in its use in all situations. Similarly, when talking about use we cannot exclude the technological dimension.

In order to elucidate the ties between the frame of functioning and the frame of use, we need to examine their articulation more precisely. First, it seems that the designer considers use on two different levels: abstract use, and social use. When Alexander Graham Bell decided in 1875 to abandon the multiplex telegraph project (simultaneous transmission of several messages) on which he was working, to devote himself to telephony, he conceived of a new technical use for the telegraph network: the transmission of sound. But a number of possible social uses corresponded to this abstract use: transmission of music, transmission of voice messages, conversation over a distance. The definition of abstract use and that of social use are two distinct issues. One concerns the frame of functioning, the other the frame of use. The definition of the (semaphore) telegraph in the dictionary of the French Academy in 1837 (*Nouveau vocabulaire de la langue Française*, 1837: 626) clearly illustrates these distinctions: 'Newly invented machine which by means of different signals [frame of functioning] transmits over great distances and in a very short space of time [abstract use] everything which might be of interest to the government [social use]'. Finally, the designer has to answer three types of question: which use, associated with the frame of functioning; why this use, related to the frame of use; and how it is used, relating to interfaces.

To specify the necessity of the separation between the frame of functioning and the frame of use, we shall pause for a moment on the problem of a breakdown. All of a sudden the machine 'which works alone' can no longer be used. The question is no longer one of the task to accomplish or of the pleasure obtained from its use. The machine has to be repaired and in order to do so the black box must be opened. The user then enters into another sphere of reference. There is a shift from the frame of use to the frame of functioning. The buyer of an object in kit form is faced with a similar but reverse situation. Before the purchase can be used and enjoyed, it has to be assembled. It is therefore necessary to take care of the frame of functioning before the frame of use.

Nevertheless, the break between the two frames is not complete, for they are articulated in a common framework, that of the motor car and that of micro computing. This brings us to the well-known approach of Thomas Hughes (1983) who presented Edison's incandescent lamp as a 'seamless web' combining technical, social, economic and other elements. The link between the two frames appears clearly in exceptional situations where one of the two evolves while the other one remains stable. Take, for example, the successive transformations of the calculator. At first the change in the frame of functioning (shift from electro-mechanics to electronics) did not alter the use. The keys of the first calculators were almost identical to those of electro-mechanical machines. Later, the power and rapidity of electronics opened new possibilities for calculation and the frame of use was modified.

Thus, frames of reference can change. A car which is normally used to travel in can also be put at the bottom of a garden to serve as a chicken shed, or in a street to serve as a barricade. Ronald Kline and Trevor Pinch (1996) studied a far more interesting case of an alternative frame of use. In contrast with the car as a means of transport characteristic of an urban model, we find an alternative frame in the rural world, that of a farm tool, a stationary source of power (for a water pump, wood saw and so on). Hence, in the first quarter of the twentieth century we witness a controversy over the frame of use.[4]

This change in the frame of use is not entirely independent of the frame of functioning. In the first case it is probably because the car no longer functions that it becomes a mere dwelling place. In the second case it is precisely because it has wheels that it can be pushed into the middle of the road, and because it has a petrol tank that it can become an instrument of urban guerrilla warfare. In the case studied by Kline and Pinch, farmers initially tinkered with the well-known Ford T, and afterwards manufacturers produced and commercialized an accessory: a pulley to be attached to a jacked-up wheel. But they realized that jacking up one wheel put undue strain on the differential gear and so found a new solution (ibid.: 786). Thus, in this case, the definition of a new frame of use altered the frame of functioning.

Such transformations in the frame of reference have the advantage, for the observer, of highlighting the existence of a frame of coordination for individual technological actions. The change of a frame of reference is also part of the dynamics of technological evolution. There may therefore be changes in the frame of functioning and in the frame of use. These two movements have their own dynamics and are mutually interactive.

Under no circumstances do frames of reference determine technological action. Rather, they provide a point of anchorage, a set of constraints which make the technological activity possible. The latter takes place freely within this framework.

We shall now return to Goffman and examine face-to-face communicational interaction. Goffman distinguished system constraints and ritual constraints. The former correspond to 'conditions and devices' making it possible to facilitate the efficient transmission of speech. The latter concern 'how each individual ought to handle himself with respect to each of the others' (Goffman, 1981: 16). Michel de Fornel extended these Goffmanian problematics to mediated communicational interaction and particularly to the case of the videophone. To the system constraints which, with Goffman, depend on the physical and cognitive characteristics of the individuals involved, he added other types of constraint related to the technological object (de Fornel, 1988, 1996). The notion of a frame of reference is situated in the same perspective, but it extends the question of articulation between the technical and the social to situations affecting usage as well as design.

Robinson used the concept of a 'common artifact' to denote technological devices whereby cooperation between individuals could be organized. In particular, he studied two such devices. The key-rack in a hotel makes it possible to know which rooms are available and whether the guests are in, to distribute messages to them, and so forth. In air traffic control, flight progress 'strips' are instruments of cooperation used by air traffic controllers. Common artifacts have the following characteristics: they are predictable, allow for rapid information and implicit communication (Robinson, 1993: 190–95) and constitute the information framework of cooperation. In this case, the frame of functioning appears as the medium and instrument of cooperation.

In the conclusion to his article on the videophone, de Fornel wrote: 'It seems that the sociology of use, like the analysis of interactions, has tended to perceive the technological object as a "black box"'. Research on usage 'tends to maintain, implicitly or explicitly, that social uses are in no way inscribed in the instruments themselves'. By contrast, 'studies of mediated interaction tend to see in the media nothing more than a set of technical constraints' (de Fornel, 1988: 45). In short, 'all these analyses have the shortcoming of largely failing to consider the status of the technological object in relation to communicational activity' (ibid.: 46).

Louis Quéré, who belongs to the same French ethnomethodological movement, raised the same key question of a study of technology in everyday life: 'How do we manage to hybridize technological objects on the one hand, and social practices, beliefs, values and norms on the other?' (Quéré, 1992: 31). His theory was then that 'objects are endowed with an "innerness" (totally independent of the functioning which defines the technological object) through their incorporation in our social practices and through the integration, in them, of our capacities, our uses and the symbolic systems which mediate our practices' (ibid.: 32). To define this inner nature of the technological object, Quéré drew upon the work of Georges Mead. The latter considered that machines ought to be treated not as objects endowed with their own qualities, but as entities which incorporate certain characteristics of usage. Mead described the interaction between an object and the organism that used it: 'The formula for this process is that the thing stimulates the organism to act as the thing acts upon the organism, and that the action of the thing is the organism's resistance to pressure such as arises when a hard object is firmly grasped in the hand. The resistance of the object is continuous with the effort of the hand' (Mead, 1932: 122).

The fact that a pragmatist sociologist like Mead took an interest in technological objects is a point worth emphasizing. He was thus one of the first sociologists to have fully understood that the object is not outside the social, it is at the heart of our societies. The social link is mediated by the object.

Quéré extended his reflection, starting with an analogy with the question of events. How does an undefined occurrence suddenly acquire social meaning, become incorporated into a network of shared references and give rise to specific reactions? This view of the event can be extended to any technological action. Quéré was primarily interested in the questions of use and technological discourse, but if we broaden this perspective to the design of machines we find the notion of a frame of reference.

In short, the socio-technical frame of reference makes it possible to perceive and understand the technological phenomena that we witness, and to organize our action and our co-operation with other actors. It comprises a body of knowledge, know-how and technological artifacts mobilized during the course of technological action. By means of the frame of reference the interaction which an individual develops with technological artifacts and with other people can be structured. It organizes an individual's interpretations and deliberations with him- or herself.

This notion enables us to avoid the dead-ends of a sociology of use which refuses to 'open the black box', or those of the Actor Network Theory which sees use as nothing but an undefined field of socio-technical networks. The notion of a frame of functioning furthermore indicates that the flexibility of networks is not infinite, and that not all socio-technical combinations are possible.

SOCIO-TECHNICAL ACTION AND INNOVATION

Having examined the constitution of frames of reference and particularly the combination of the frame of functioning and the frame of use, we shall now turn to the activity of the actors of the technology (that is, the actors involved in its design, creation and use). This subject will be considered in two steps: first within the period of innovation and development of the socio-technical frame, and then in the period of stability, that of ordinary technological action.

Two groups of actors of technology can be distinguished: strategists and tacticians. I have borrowed this distinction from Michel de Certeau but have modified its application since for him it opposes writing and reading, urban planning and city dwellers and, more broadly, production and consumption. De Certeau's distinction is inspired by the one which linguists make between 'language' (a system) and 'speech' (an act): 'The act of saying is a use *of* language and an operation *on* it' (de Certeau, 1980: 82; original emphasis). He then changes his referent by moving from linguistics to polemics.

> [Strategy] postulates *a place* likely to be circumscribed as a 'home base' from which relations with *an exteriority* can be managed ... The 'home base' is a *victory of the place over time*. It makes it possible to capitalize on acquired advantages, to prepare future expansion and thereby to have independence in spite of variable circumstances. It is a control of time through the creation of an autonomous place. (Ibid.: 85; original emphasis)

In contrast, tactics are characterized by 'the absence of a home base. Their only place is that of the other. They therefore have to play with the field imposed on them' (ibid.: 86). They constitute the art of taking advantage of opportunities.

These two concepts of strategy and tactics enable us to distinguish two types of actor: those who participate in the creation of a frame of reference and those who are subjected to it. More profoundly, this strategy–tactics distinction allows us to review the question of relationships between intentionality, opportunity and context. Bruno Latour, for example, in *Science in Action*, uses the metaphor of Scrabble to present the innovator. Just as a good player is not one who prepares his words in advance, so too a scientist or engineer must above all grasp opportunities. I think, on the contrary, that technological actors first define a project and then acquire the means to realize it in a space which they delimit. In this position they are strategists. As soon as they move out of this space they enter into a context which they no longer control. Opportunities arise, of which they take advantage or not, and in this position they are tacticians.[5] During the phase in which frames of reference are developed, a single actor may alternate the roles of strategist and tactician. For designers, the boundary between strategy and tactics is not the one separating their laboratory from the outside world; it runs through their laboratory. They may count on a particular component and if it proves unreliable they will have to choose another one. Thus, the frames of reference of the different technological artifacts fit into and overlap with one another. For example, the designers of a new type of telecommunications network will use components or cables which may belong to other frames of reference, themselves in gestation. When Bell and Elisha Gray launched their respective work on the multiplex telegraph, they each developed a strategy. Both of them were faced with an opportunity, that of transmitting sound. While Bell grasped the opportunity, Gray did not do so immediately. Bell's tactical skill is unquestionable, but he would never have been able to implement it without a strategy (Hounshell, 1975).

Finally, we note that the higher upstream one is in the history of a technological artifact, the greater the interplay between strategy and tactics will be. As we move further downstream, tactics increasingly prevail. When the frame of reference is completely established, technological action is entirely tactical, unless the frame of reference becomes obsolete and is replaced by another one, in which case strategic-type action will of course become necessary.

In the strategic actions which make up a frame of reference, it is important to bear in mind that a number of principles have to be defined in order to make cooperation between very different actors possible. In terms used by interactionist sociology, a boundary frame has to be defined, which must then be adopted by the laboratory, factory, sellers, repairers or users.

The advantage of the concept of a strategy is to show that a frame of reference, unlike a language, is not a system established once and for all. On the contrary, it is constructed by individual actors. Other actors are then tacticians; they act in a predetermined frame of reference and use all opportunities to develop specific technological objects. Thus, when Henry Ford developed the motor car for a mass market, he developed not only a specific technological object, the Model T, but also a type of division of labour – the assembly line – and remuneration of labour. General Motors (GM) then diversified the models, adapting both production and use (Hounshell, 1984). In the Fordist system we can consider that Ford is a strategist and GM managers are tacticians. On the other hand, if we focus rather on the question of the range of cars, we can consider that GM managers become strategists. Hence the strategy–tactics distinction has meaning only in relation to a specific type of socio-technical action.

Like Ford, the strategist influences the frame of functioning, the frame of use, and their articulation. Consider another example, the dispute between Thomas Edison and Auguste and Louis Lumière at the time of the birth of the cinema. Initially Edison conceived of a device similar to the phonograph, with images placed in spirals on a cylinder, intended for individual viewing. He later adopted the solution of perforated film. The first commercial use proposed by him was viewing in slot machines. The content was very simple: a few characters on a black background. Lumière, on the other hand, proposed a projector and his films were made outside and designed to show movement. On that score Lumière triumphed. But the socio-technical frame of the cinema, in the form in which it was adopted, combined elements developed by Edison, Lumière and many others (Flichy, 1995).

While, in a sense, the frame of reference combines the action of rival designers, it constitutes, *a fortiori*, the framework of cooperation of different actors in the same laboratory of the same organization. R&D projects unite diverse knowledge and expertise which most often belongs to different teams in the same institution. Cooperation between these teams is complex and can only take place if, following conflict and negotiation, they agree on a common boundary object. In the example of the videotext, development of the technology required collaboration between designers of the terminal (the Minitel), database hosts, two networks (the telephone network and then Transpac) and the interface between these two networks provided by videotext access points. These access points also supplied the reception pages that users saw on their screen. All these teams were able to cooperate only because they had created a common frame of reference.

The notion of designer-strategists also enables us to explain the fact that innovation is not merely the fruit of chance, and that all firms do not have the same assets in technological development. We cannot consider that a given socio-technical system is determined accidentally. Rather, we need to see that strategic actors can play a key role. Christopher Freeman, in the context of the Sappho project (Freeman, 1982: 109–15), showed that those firms which were able to lay the foundations of an innovation strategy by means of efficient R&D teams and a thorough understanding of their current market, secured a competitive advantage.

Beside these designer-strategists there may be user-strategists. It is they who can specify the functionalities of the technological tools they wish to use. Usually, they are firms which negotiate with the designers within a formal framework. In the case of nuclear power stations or large telecommunication systems, the builder's laboratories will negotiate with the network operators who, in turn, have their own laboratories. The latter may build mock-ups of the desired equipment or at least define functionalities in detail. The same applies to the field of scientific instruments, where the frame of reference is developed together by user laboratories and manufacturers. A survey conducted by Eric von Hippel (von Hippel, 1988) on over a hundred innovations in the field of scientific instruments, showed that three-quarters of them were developed on the initiative of users. In this book, von Hippel associates these user-strategists with 'lead users'. They perceive general needs before the marketplace encounters them, and are positioned to benefit significantly by obtaining a solution to those needs. Von Hippel generalizes the notion of the user-strategist beyond cases of instruments and networks, to the case of the general public. Final users could also be in a strategic position, for instance in open source software. But this is not the case for everyone; many are mainly tactician innovators. The example of high-performance windsurfing that von Hippel gives at the beginning of a recent book clearly illustrates this kind of innovation (von Hippel, 2005: 1).

In other cases users do not negotiate the frame of reference *ex ante* but transform or even impose a frame of use *ex post*. That was, for example, the case of bankers in the 1830s who tried to use the semaphore telegraph to transmit financial information. They challenged the frame of reference of the telegraph as defined after the French Revolution: transmitting state messages (Flichy, 1995). The first American firms that installed computers in their administrative services at the end of the 1940s, while these machines were primarily intended for scientific calculations, were also user-innovators (Cortada, 1993).

Individual users may also form pressure groups in order to modify a frame of use. That was the case in 1993 when French user associations obtained changes to Socrates, the system of train reservations. More generally, they also obtained a return to the public service type of use as well as the rejection of a proposed new framework which was to resemble that of flight reservations. Nevertheless, this situation of user-strategists is exceptional when it concerns the public at large. Usually designers negotiate with a virtual public. They are faced not with users themselves but with their own conception of them.

Finally, there are cases in which designers and users interact as much on the frame of functioning as on the frame of use. JoAnne Yates, who studied these interactions between tabulating industries and life insurance companies in the early twentieth century, concluded that 'the insurance industry was not a passive, if eager, recipient of this technology'. As lead users, these companies 'encouraged certain types of development in the technology' (Yates, 1993: 48). On the other hand, the new technical configuration altered insurance work by automating certain tasks, reducing errors and so on.

EVERYDAY SOCIO-TECHNICAL ACTION

When the socio-technical frame is stabilized, the different actors' activities vary considerably. They have to fit into the frame and are therefore tacticians. We shall first consider the situation of designers. For each specific technological object, multiple choices have to be made. However, certain alternatives to the frame of reference are not examined since the designers do not naturally think of them. A frame of reference defines questions to be studied (in terms of functioning and use) and developments and improvements to be made. The historian Michael Baxandall considered the case of the designer of a bridge in the nineteenth century, where the project was to build a viaduct over the estuary of the river Forth, to link Scotland and England (Baxandall, 1985: 12–40). The frame of reference was given, it was that of a metallic bridge for a railway line. The engineer responsible for the construction, Benjamin Baker, had to accomplish a specific task, that of crossing an estuary which had particular characteristics concerning its banks, depth, bed and so on. The bridge had to withstand strong winds, a condition which was reputedly difficult to meet since another Scottish bridge, over the Tay, had been destroyed during a storm. Finally, the bridge had to carry trains of a given weight. Faced with this problem, the frame of reference (Baxandall speaks of 'technological means' and 'technological culture') offered different options to Baker: laminated iron or steel, bridge suspended with chains or self-supporting girders. Not only did the engineer choose between these different technological solutions, he also widened the range of possibilities by building a cantilever steel bridge. This example shows that the frame of reference neither determines the technological object nor completes its description. To understand the specificity of the artifact and to reconstruct the multiple choices made by the engineer, we constantly have to examine the interaction between the relevant characteristics and the frame of reference. In Baxandall's words, these are 'the terms of the problem' and 'the culture'.

Numerous other actors are involved in the frame of reference. In mass production the methods department has to define the organization of production in detail. In relation to the overall frame of reference of a technological artifact, this department is in a tactical position. However, in relation to the operators in front of their machines, it is in a strategic position since it defines the framework of their activities. Operators are in a tactical position on all levels, in relation to both the product being manufactured and the machines they use. Maintenance technicians are likewise in a tactical position. Even if they are able to use clever tricks of the trade to get round the frame of functioning, they are always limited by a frame.

Users' Action

With the exception of some rare user-strategists who may intervene in the creation of the frame of reference, individual users are, in most cases, tacticians. We find here de Certeau's problematics concerning the 'art of doing' and styles of action. These 'intervene in a field which places them on a first level, but they introduce a way of taking advantage of the situation which adheres to other rules and constitutes a sort of second level fitting into the first' (de Certeau, 1980: 76). This operative art consists, for users, of acting otherwise than in the way defined a priori by the frame of use. De Certeau, notes Luce Giard, shows 'that there is creativity of ordinary people. It is hidden in a confusion of silent, subtle and effective tricks whereby everyone invents their own way of moving through the forest of products imposed on us' (Giard, 1990: xxiii).

The activity of the user-tactician constitutes a kind of poaching. It is faced with the frame of functioning that users can first show their tactical skills, that is, faced with a machine which imposes its operating procedures and offers resistance. As Gilbert Simondon notes, the machine incorporates the experience of its designers: 'Through technological activity man creates mediation, and this mediation is detachable from the individual who conceives and produces it; the individual expresses himself through it but does not adhere to it; the machine possesses a sort of impersonality which allows it to become an instrument for another man' (Simondon, 1989: 245). Thus, it incorporates a specific form of functioning: 'The user has to possess forms within himself so that, from the encounter between these technological forms and the forms conveyed by the machine and realized within it, meaning arises from which work on a technological object becomes technological activity' (ibid.: 250). According to Simondon, this compatibility of forms is possible only if the machine is open enough for the user to extend the manufacturer's act by fine-tuning, setting and repairing it. The user must then have an adequate technological culture.

We can, it seems, apply Simondon's thought to a different context from that of production tools, one which is adapted to situations in which users' knowledge is more limited. In such a context, as Jouët states, 'the user-culture is thereby gaining being technical features which do not of course constitute a technical culture as such, but which are gradually permeating people's habitual frames of reference' (Jouët, 1994: 75).

In all instances, the machine incorporates the user's project: to watch a video, warm a drink, type a text and so forth. The user knows what can be asked of the machine and an adjustment between the two can be made well within the limits of its technological performance. Thus, while most people who have a VCR pre-set it to record a television programme at a particular time, a significant minority slides the tape into the machine at the scheduled time (or just beforehand) (Arnal and Busson, 1993: 142–50). Despite a lack of know-how, this minority has found a way to adjust to the machine.

Numerous electronic household appliances are equipped with interfaces inspired by computing, with display and programming functions: 'The principles of programming and sequential logic are henceforth inscribed in the operation of everyday appliances and, through experience, have become an integral part of the mental schema of a large number of users' (Jouët, 1994: 74). Jouët also identifies this isomorphism of technical and social forms in the field of convivial telematic message services (Jouët, 1991). The communication software which manages dialogue between the different Minitel users structures conversation to such a degree that Jouët uses the term 'technical speaker'. In messaging, dialogue is always dual since it is conducted with the machine and with a partner (or partners) simultaneously: 'Social organization and technical syntax are the result of a close interlinking' (Jouët, 1994: 81). Julia Velkovska reaches a similar conclusion in her research on webchats. She studies the link between the space–time framework of the communication device and the form of interactions taking place (Velkovska, 2002).

Although the action of a user-tactician is situated in a socio-technical frame of reference which defines the suitability of forms between the user and the machine (discussed by Simondon and Jouët), there are nevertheless fairly broad margins of leeway enabling everyone to use a technological tool in their own way.

These tactical capabilities of the user are obviously far greater when it comes to the frame of use. In the 1950s, teenagers took their parents' new wirelesses from the family living room to their own rooms to listen to rock music. They thus transformed the use of the radio from a family instrument into an individual one. Similarly, TV viewers in the 1980s, equipped with a remote control device, created their own television programmes by surfing from one channel to another (Bertrand et al., 1988: 56), and users of word processors create their own mode of writing by activating specific functions of the software (Proulx, 1988).

During the last three decades of the twentieth century when investments in R&D and production reached considerable proportions, the suppliers of new technological systems often ran experiments to define pilot users before marketing their products. These first users, such as beta testers of video games (Castronova, 2001), developed tactical skills and some even bypassed the system. The results of their actions were measured by marketing people who became new strategists capable of negotiating the socio-technical frame with the designers. Thus, users did not intervene directly as strategic actors. Firms set up actors in their marketing departments who, through experiments as well as surveys on more standard practices, were capable of integrating users' tactics into a future socio-technical frame.

Having examined users' adjustment to technological objects, let us consider the question of cooperation with other actors. In order to do so I shall draw upon certain ideas from the theory of conventions. This theory postulates that 'agreement between individuals, even when it is limited to a commercial contract, is not possible without a common framework, a constitutive convention' (Dupuy et al., 1989: 142; Rallet, 1993). When coordination between actions has to take place within a small group of actors, common experiences or views make it possible to identify other actors' models without too much difficulty. By contrast, when the number of actors is indefinite and they are situated in different local situations, coordination of what Pierre Livet and Laurent Thévenot call 'action together' is provided by 'conventional objects' that incorporate common knowledge. In the domain of technological objects, these are operating manuals, plans, protocols and so on, which serve primarily as references in case of coordination problems. Nevertheless, a large number of conventions are not explicit (Livet and Thévenot, 1991).

In my view it is significant that such conventions impact on both the functioning of the technological object and its use. Take the telephone, for example. A convention of functioning distinguishes two different types of signal indicating whether the line is free or engaged. The convention of use is to introduce oneself to one's interlocutor. By contrast, with digital telephone terminals on which the caller's name is displayed, that social convention is altered, so that interlocutors no longer have to introduce themselves. Thus, with a communicational artifact such as the telephone, relational adjustment may be provided by a system of functioning or by a system of use, with the balance between functioning and use varying from one machine to another.

The research undertaken by de Fornel on the videophone demonstrates the role of the technological object in the coordination of communicational actions. Videophonic conversation is situated between telephonic and face-to-face conversation. Users who have not yet integrated communicational conventions of the videophone tend to take as a reference the medium with which they are already familiar – the telephone. Instead of videophoning, they watch each other telephoning. Those who wish to exploit the resources of the image have to learn new communicative skills. Like any social individual, they have learned at what distance from the other person they should place themselves and what pitch of voice and gestures they should use in face-to-face conversation. They apply this knowledge 'in relation to a complex body of criteria, such as the environmental and social context, the degree of familiarity with the other person' and so forth. In a videophonic conversation they have to incorporate, in existing conventions, 'the system constraints peculiar to the technological object (positioning in relation to the camera, determining the appropriate distance from it, using the control function, etc.)' (de Fornel, 1991: 126).

It is in the opening of conversation that the evolution of conventions appears most clearly. Users start with a telephonic opening ('hello' and then introduce themselves). They then add a new series of greetings after having switched on the image. Experienced users, by contrast, fully utilize the possibility of visual identification, which necessitates switching on the image immediately. However, even though the videophone generates specific conventions of sociability which integrate certain technological elements, each user still develops his or her own videophonic style. Users mobilize the expressive resources of the face when their image is close up, or personal gestures if they are further from the camera.

This example of the videophone highlights the main components of the socio-technical frame of reference approach. This frame is the element of stability that allows for cooperation to develop between the technical actors, on the one hand, and between the actors and the technical artifact, on the other. It is also the outcome of a construction by actor-strategists and a re-appropriation by actor-tacticians who develop specific technical styles in the technical operation of the device and in its use. The socio-technical frame organizes the collective activity and allows for some individual differentiation. It offers a basic structure on which everyone can improvise:

designers who differentiate their product from those which already exist; users who differentiate themselves from other users.

NOTES

1. Ethnomethodology can be defined as 'the empirical study of the methods that individuals use to give meaning to and to accomplish their everyday actions, that is, to communicate, make decisions and reason' (Coulon, 1993: 26).
2. For a presentation of Garfinkel's experiment and an overview of ethnomethodological research in the scientific field, see Lynch et al. (1983: 205–38). See also Lynch (1993).
3. This text includes an accurate presentation of the interactionist approach to science.
4. The sociology of techniques, which has paid a great deal of attention to technological controversies, has focused more on controversies over the frame of functioning than on those over the frame of use. Another interesting case of controversy over the frame of use is found with the early twentieth-century telephone. Whereas the engineers who designed the telephone conceived of it as a tool for sending messages, rural users turned it into an instrument for social conversation over long distances (Fischer, 1992b; Flichy, 1995).
5. In Suchman's (1987) terminology, we could say that forecasting action belongs to strategy while action belongs to tactics, or that strategy is a resource for tactics.

PART III

Socio-technical history

5. The time of technology

In the case of individuals who undertake technological actions, the frame of reference is stable. By contrast, when we consider a different time-scale, not ordinary socio-technical action but the emergence of an innovation in technology or usage, that frame is mobile. We then need to consider the birth of a socio-technical frame that we know will subsequently remain relatively stable until it is replaced. As Laurent Thévenot notes, 'the differentiation of regimes of adjustment enable us to distinguish the point along the way on which economists of path-dependency focus, from the point at which judgements are made with reference to set conventions. The latter implies detaching the convention from the contingency of its history' (Thévenot, 1993: 141). Whereas at the end of the previous chapter I examined socio-technical action in a context of 'set conventions', in this chapter I study the birth of those conventions.

By switching from socio-technical action to the construction of frames of reference, it is not only our field of observation that changes (from ordinary interactions to an innovative set) but also our methods of analysis. As Paul Ricoeur notes, there is a big difference between the theory of action and the theory of history. Whereas the language of action, which combines action and intention, is independent of the outcome of the action, that is not the case of the historical account: 'In so far as putting the past into a temporal perspective underscores unintended consequences, history tends to weaken the intentional accent of action' (Ricoeur, 1991: 262). As Arthur Danto notes, 'for the whole point of history is *not* to know about actions as witnesses might, but as historians do, in connection with later events and as parts of temporal wholes' (Danto, 1965: 183; original emphasis).

Like any other account, the historical account is governed by the way it ends. It is therefore a conception of time that differs from that of the actors for whom time flows from the past towards the future: 'It is as if the recollection reversed the so-called "natural" order of time. ... We learn to read time itself backwards, like the recapitulation of the initial conditions of a course of action in its terminal conditions' (Ricoeur, 1991: 131).

The historian's work consists in repositioning different events in relation to one another, to construct a meaningful whole. Paul Veyne considered, more precisely, that historical explanation is constructed along the three lines of chance, material cause and freedom (Veyne, 1971: 120–21). Once complete, the historian's account could nevertheless produce the 'retrospective illusion of fatality' referred to by Raymond Aron (1957: 187). To avoid this stumbling block and 'restore to the past the uncertainty of the future' (ibid.: 182), it is necessary to construct an alternative course of events, via the *imaginaire*, and to compare that to the real one. This method, which Max Weber calls 'causal imputation', enables the historian to 'grasp the relationship between the essential elements of the event and certain elements caught in the infinity of decisive moments' (Weber, 1965: 300). For Weber, the application of this method is a key element in the historian's work.

> [If the historian] communicates to the reader, in the form of an account, the logical result of his historical judgements of causality, without giving details on his sources of information … his account is nothing more than an historical novel. It is not a scientific relationship if the sound framework of causal imputation is lacking in the external arrangement of the artistic presentation. (Ibid.: 307)

The Weberian method leads us to causality in history. In Chapter 2, I presented historical works that defended the thesis of a causal link between technology and social organization. These works generated stormy debate, especially since some of them were based on a notion of causality derived from the natural sciences, in which regularity is associated with causality: whenever an event A occurs at a certain time in a certain place, an event B occurs at an associated time and place. This nomological view of causality was opposed by another view held by a current of epistemology of history. Ricoeur considered that the historian seeks to ascertain 'in which respects the considered events and their circumstances differ from that with which it would be natural to group them under a classificatory term … The historian proceeds not from the classificatory term towards the general law, but from the classificatory term towards the explanation of differences'. But this explanation, says Ricoeur, is

> [a matter of] judgement rather than deduction. By judgement I mean the sort of operation performed by a judge when he or she weighs up contradictory arguments and takes a decision. Likewise, for a historian, explaining means defending his or her conclusions against an opponent who invokes another set of factors to support a thesis. This way of judging on particular cases consists not in putting a case under a law, but in grouping together scattered factors and in assessing their respective importance in the production of the final result. Here the historian follows the logic of the practical choice rather than that of scientific deduction. (Ricoeur, 1991: 223)

The socio-technical history to which I refer in the rest of this book is not the traditional history of dates of inventions. In the perspective of the

École des Annales, I am as interested in micro history, that of an invention or a laboratory, as in the long periods so dear to Fernand Braudel, that is, technological paradigms or mentalities. But this 'immobile history' becomes history only when it takes changes into account. Just as any socio-technical action is situated in a frame of reference, so too the historical approach of interest to me is that which proceeds from the precise study of the inventor's work to the study of the major technological and social trends that structure frames of reference.

This history of frames of reference will be presented in three stages. First, in this chapter I examine how economist-historians incorporate time into their analyses of technology. In the following chapter my perspective is that of the history of mentalities, as I study how the creation of frames of reference is articulated to the *imaginaires* of an era. Finally, in Chapter 7, I combine these different elements and propose a unified approach to the birth of frames of reference.

Although in the present chapter I have selected three analyses of time corresponding to different economic approaches, I also make several incursions into the sphere of historians and philosophers.[1] I first consider the way in which certain economists study technological choices to explain how a frame of reference is set or, to use the vocabulary of the new sociology of technoscience, how we go from a controversy to a black box. After considering this question of the choice of a technology studied over a short period, I examine more long-term phenomena. How is technological experience built up? How does it evolve, step by step? These are the first questions to answer, but there are others. How do big changes occur? How do we switch from one frame of reference or one paradigm to another?

TECHNOLOGICAL TIMES AND CHOICES

In Chapter 1 we saw that in standard neo-classical analysis, technology is an exogenous factor that hardly interests economists. By contrast, classical economists incorporate technology into their analyses. Nevertheless, the two currents do agree on one point: technology evolves continuously through a series of minor alterations or improvements. Whether we consider it as endogenous or exogenous, technological progress is an essential factor of economic growth. Even if, in Karl Marx's theses, economic and social trends are constantly faced with contradictions, conflicts and struggles, the dynamics of productive forces (especially those of technology) are continuously growing. Alfred Marshall, founder of the neo-classical current,[2] likewise considered technological change to be of an incremental nature and therefore continuous.

Alongside the 'standard' theoreticians there have been unorthodox economists who have insisted on the discontinuities of technological progress. Joseph Schumpeter is probably the best known. Further on in this chapter I consider the production of his contemporary disciples and, more generally, the evolutionary current that studied both technological trajectories and technological change. Another group of contemporary economists has considered technological evolution to be so discontinuous and unnatural that it has placed the question of technological choice at the heart of its research. This current, consisting of micro economists, believes that the success of one technique in relation to another cannot be explained only in terms of the 'invisible hand' of the market. Unlike the dominant economic approach that sees the individual (whether producer or consumer) as facing his or her environment alone, the technological competition school focuses on interactions between the agents, who know only specific, local aspects of the economy concerning themselves directly. Adjustment and coordination between the agents takes place at this local level. In this conception, economic agents' technological choice depends less on the global state of the market than on the position of their reference group. In relation to the concepts developed in the preceding chapter, we can say that technological activity always takes place in a framework of reference. That framework is not universal, it is peculiar to each social group. The higher upstream one is in technological development, the more micro groups (research teams) there are constructing distinct frameworks of reference. When technological innovations arrive in the market, some of these frames clash, and more global balances appear. However, segmentation of markets at a geographic or social level causes technological choices not to become universal. Various techniques and standards can coexist.

This analysis of technological choices with its combination of decentralized choices and interactions between agents, questions not only the perspective of a universal, optimal choice provided by the invisible hand of the market, but also the idea shared by most engineers and scientists that at a given point in time, one technique and only one will be optimal. In other words, the preceding theory challenges not only market but also technological rationality. There is no natural and universal technological solution but various solutions, constructed in the complex play of interactions. As Dominique Foray wrote, 'we don't choose a technology because it is more effective; it is because we choose it that it becomes more effective' (Foray, 1989: 16). Brian Arthur, one of the leading US researchers who identifies with this current, considers that five factors explain the growing appeal of a new technology when its diffusion increases:

1. *Learning by using*[3] The more a technique is used, the better people learn to use it, the more its use is optimized, and the more efficient it becomes. Nathan Rosenberg, who was the first to signal this effect, noted in respect of the aeronautics industry: 'during the operation of a new aircraft, operating cost reductions depend heavily upon learning more about the performance characteristics of the system and components and therefore upon understanding more clearly the full potential of a new design' (Rosenberg, 1982c: 125). What is more, learning makes it possible to discover not only new technical potentialities, but also new uses. Initially the telephone was used essentially to transmit messages. It was only later that it became a tool for conversation, for distance visits (see Martin, 1991; Fischer, 1992b).

   Finally, Foray wrote:

   > [L]earning by using gradually changes the way in which a technology is evaluated, compared to rival technologies. A new technology is always evaluated by means of performance criteria peculiar to existing technology, for which it is supposed to be a substitute. In other words, old technology imposes on new technology its own norms of economic evolution, formulated in relation to its natural qualities, thus introducing a sort of bias at the time the economic calculation is made. ... The introduction of compound materials in the motor car was also hindered by its poor weldability. This is a criterion of comparison that is valid only in the context of 'metallurgical' design of the vehicle. (Foray, 1992: 62)

2. *Network externalities* This economic mechanism refers to the fact that the utility of a technique increases as the number of adopters rises. By definition, this is the case of network technologies such as telecommunications. The usefulness of the telephone and Internet, for instance, increases along with the number of subscribers to the services. In a situation of competition between two telephone networks in the same geographic area, which are not interconnected, competition between the two companies will affect all potential users. This space is measured not only in quantitative but also in qualitative terms. Bell Canada was fully aware of that when in 1880 it offered all doctors in Montreal a special phone line. Soon subscribers of the rival operator switched to the Bell Canada network because they wanted to be able to phone their doctor (Martin, 1991).[4] The Internet is another fine example of network externalities. During the 1990s, it doubled in size every year. In the computer community this phenomenon is known as Metcalfe's Law, which states that the value of a network is proportional to the square of the number of users (Shapiro and Varian, 1999: 239).

Along with this effect of direct externality we also find indirect effects. For example, the more cars there are with a new type of engine, the more outlets there will be for the new fuel it uses. Likewise, the more a standard VCR is sold, the easier it is to find unrecorded or pre-recorded tapes for that standard.

3. *Economies of scale, learning by doing* A technology that is widely diffused has to be mass produced. Specific production devices that allow significant economies of scale can then be developed. To this phenomenon can be added learning by doing. In this case reference is no longer made to increasing quantities but to the gradual accumulation of production experience that makes it possible to reduce costs. All these phenomena have been studied for a long time by economists.[5] The originality of the new school of technological change is that it shows that these transformations in conditions of production alter the potential diffusion of the new technique. Unlike the theories discussed in the first chapter, there is no longer a break here between production and diffusion.

4. *Increasing returns on information* The more widely a technology is diffused, the better it will be known. The fear of having all the initial problems to put up with disappears, and the circulation of information helps to extend diffusion. This coincides with economic and sociological theories on technological diffusion. Some researchers of the school of technological change, like Robin Cowan (Cowan, 1990), have refined this analysis. They explain that the profitability of a technology is unknown or, at least, badly known at the outset. During its development, uncertainties on profits die away. This is a new factor boosting diffusion.

5. *Technological complementarities* 'Inventions hardy ever function in isolation. Time and again in the history of American technology, it has happened that the productivity of a given invention has turned on the question of the availability of complementary technologies,' (Rosenberg, 1982a: 56). With historians of technology we find many examples of this articulation between techniques. Rosenberg cited the case of electricity, which Edison saw as a real system, and Thomas Hughes shows, in *Networks of Power* (Hughes, 1983), how a technical network necessarily intersects with larger social, political and economic networks. For instance, electrical systems include not only utility companies, but also manufacturing firms and investment banks. Bertrand Gille contrasted the example of 'technological branches', consisting of the different products successively involved in the production process, and 'technological systems', in which coherences are far more complex and interactive: 'If the iron and steel industry uses the steam engine, that engine needs increasingly strong metals to withstand high pressures and

overheating' (Gille, 1978: 18). Hence, the steam engine and the iron and steel industry are part of the same technological system.

In Arthur's perspective, two points can be noted on the subject of technological complementarities. First, the more a technique is adopted, the more it will attract related techniques that make it more attractive. Moreover, the development of a technique can depend on outside events that boost its growth. This is very important. Although in my presentation of the above points I emphasized the aspects of self-reinforcement of a technique during its development, it is important to bear in mind that there are also outside events that impact on that course.

The five mechanisms described above make up what Arthur called 'increasing returns to adoption' (Arthur, 1988: 590). And it was precisely with this hypothesis of increasing returns, of dynamics of techniques, that he distinguished himself from the economic standard in which diffusion of a technology is studied only *a posteriori*. Therefore, returns can never be decreasing since the inputs required for production become rarer. In this perspective, the economist is hardly interested in the dynamics of technological choices since they are set once and for all. The school of technological change grants a fundamental role to time and to series of choices (which concern design as much as diffusion): 'In the increasing returns to adoption regime' wrote Foray, 'it is history (and not science or technology) that decides' (Foray, 1992: 66). By the accumulation of learning and the effects of network or scale, the individual decisions of economic agents induce positive feedback. The appeal of a technique and its future therefore depend on its past, on the different stages that it has gone through. This is a path-dependent economic situation.

Paul David (1986) compared this dynamic evolution to a tree. At each branch the economic agent is faced with a technological choice. Initially each choice is basically random, but once a solution has been opted for by a large number of actors, the accumulation of learning makes it relatively permanent. Choices early on in the process (close to the trunk) obviously have far more significant implications for the future than those made later on (close to the terminal branches). This process is therefore both random and deterministic. Technological competition is unpredictable – the outcome cannot be foreseen on the basis of information available at the outset – and inflexible (Foray, 1992: 74) – it occurs when one technology has finally prevailed over the others. We then refer to 'technological lock-in'. Arthur saw the articulation of two phenomena here: '(i) that choices between alternatives technologies are affected by the numbers of each alternative present in the adoption market at the time of choice ... (ii) that small events outside the model may influence the process so that a certain amount of randomness must be allowed for' (Arthur, 1988: 597). As regards the influence of preceding adoption rates, Arthur notes that the distribution of technological alternatives in the market at the time the decision is made is less important than the actors' expectations: 'Therefore if, prior to adoption, sufficient numbers of agents believe that network A will have a large share of adopters it will; but if sufficient believe B will have a large share, it will' (ibid.: 601). Thus, expectations of success tend to be self-fulfilling prophecies (Shapiro and Varian, 1999: 12).

As we have seen, this determinism–chance dialectic leads on to a situation of technological lock-in that is particularly stable when big investments have been made that would become entirely futile if an alternative were selected. But this technological equilibrium is not necessarily established at the optimum. Arthur notes: 'the economy sometimes locks in to an inferior technology because of small historical events' (1988: 595).

David (1985) was one of the first historian economists to study this phenomenon of non-optimal domination with the example of the QWERTY keyboard. This keyboard, developed in the US in the 1870s, was chosen because it ensured that the bars carrying the characters of the typewriter would not become entangled. The aim was to slow down the pace of typing. This keyboard was chosen by Remington, which altered it slightly to enable demonstrators to type the word 'typewriter' easily and quickly. All the letters of this word are on the first line of the keyboard. Here was proof that technological and commercial constraints were compatible!

With the advent of electric typewriters and then of computers, this keyboard was no longer justified in any way whatsoever. In the 1930s, August Dvorak proposed another keyboard that allowed a 20–40 per cent improvement on productivity. Despite a few attempts, the Dvorak system was never introduced.[6]

Cowan (1990) gives a similar example from the civil nuclear sector. The US opted for the pressurized water system, originally for the highly compact reactor of the nuclear submarine *Nautilus*. This option, which had already been tested on submarines, was transferred to electric nuclear power plants when the US government decided to launch a civil nuclear programme. In the Cold War context the coherence of military and civil nuclear policy seemed to be a necessity. With hindsight, however, it appeared that the pressurized water system was not the most efficient and that other solutions (natural uranium and heavy water, graphite gas) would have been more appropriate for nuclear power plants. Foray, commenting on the impact of the first order for military reactors, notes:

> [T]his event is accidental from the point of view of the dynamic process, in that it projects that process definitively towards one of its areas of extraction. On the

other hand, it is not accidental from the point of view of the historical context, in relation to which the decision taken is no doubt coherent. (Foray, 1992: 73)

This is a key point in the relationship between economics and history. When economists of technological change say that history is the decisive factor shaping technology, can they be content to consider as an accident any event other than the growth of adoption and the whole learning processes attending it? If we incorporate history into the analysis of the technological, should we not do so completely, and take the context into account as well? This is what David does in some of his work: 'In my opinion, economics cannot be a source of intellectual satisfaction unless it becomes a historical social science' (David, 1992: 242). David, who studied the introduction of the harvester in England, shows that it was adopted far more slowly than in the US, due to the characteristics of the agricultural landscape. In contrast with the wide plains of North America, the smaller fields and hilly country in England were obstacles to mechanization (David, 1971). In another article in which he considers the role of human actors in technological development, David examines the origins of a technology, when several options are available: 'It is at such junctures, I suggest, that individual economic agents do briefly hold substantial power to direct the flow of subsequent events along one path rather than another' (David, 1991: 75) and 'Some important and obtrusive features of the rich technological environment that surrounds us may be the uncalculated consequences of actions taken by heroes and scoundrels' (ibid.: 76). The importance of these positive or negative heroes who are sometimes called 'inventors' stems from the fact that there is an *ex ante* indeterminacy of technical choices and several alternatives exist. This perspective is the opposite to systemic analysis and its quasi-deterministic view: 'Once the first discoveries had been made', wrote Gille, 'everything, or almost everything, stood to reason. Distortions between the various manufacturing stages led to complementary inventions. There was thus a sort of sequence to restore destroyed equilibria' (Gille, 1978: 48).

Not only does David argue against technological determinism, but he also questions the role of human actors and thus implements the historical economics programme set by Dockès and Rosier: 'It seems that to chance and necessity we need to add a third factor, individual and collective human will which develops within multiple conflicts' (Dockès and Rosier, 1991: 197).[7] To illustrate this idea, David chose a strange example, that of the switch, in the late 1880s, from direct to alternating current. At first this example seems to invalidate his theses. In the late 1880s the networks of direct current had been so widely developed in east coast cities of the US that there could have been technological lock-in, making it impossible for alternating current to prevail. The fact that there was not lock-in is all the more surprising in the light of the fact that Edison, most historians of electricity tell us, energetically defended direct current, even going so far as to use arguments based on bad faith. Edison and his collaborators – whom some historians have called the 'West Orange Gang' – campaigned, with regard to capital punishment, to have electrocution by alternating current substituted for hanging in the State of New York. They subsequently ran a public campaign on the dangers of alternating current. In their literature, Westinghouse (the name of the manufacturer of the alternating current system) became synonymous with electrocution! (David, 1991: 89). Historians have generally concluded that even the great Edison could be blinded by his passions and refuse to see where the optimum technological solution lay, but in the end the greatest inventor of the nineteenth century was unable to stem the tide of technological progress.

David's interpretation is obviously very different. At the time, the outcome of the battle of electrical currents was not self-evident. Edison's system was composed of small urban networks of direct current of one to two kilometres in diameter around the power station. The alternating current technology offered the possibility of installing much wider networks since the transmission voltage was higher and losses were consequently reduced.

At the beginning of the controversy, the alternating current option had several drawbacks. The energy production (alternators) was inferior and the machines less able to meet peaks in consumption. No meters existed for alternating current. Finally, even though electric tramways were expanding fast, there was no electric engine for alternating current. At the same time, progress was made in direct current and the transmission of high voltage current functioned in the late 1880s not in the US but in Europe.

In the battle against alternating current, Edison did not develop the possibilities of direct current; rather he embarked on a media and legal battle. David posits that Edison had decided to withdraw from the electricity market and wanted to sell his assets in the best conditions, to develop his laboratories and reinvest in sectors with more short-term profitability: the phonograph and the cinema. His battle for direct current consisted not in developing this technical option but in discrediting the alternative, in order to be able to sell his network under favourable conditions.

The technological outcome of this controversy was the creation of a mixed system in which the direct current subsystem gradually disappeared. We thus see that actors can adopt a position not because it seems fair or effective but simply to defend or build up their industrial assets (patents, investments and so on).

To conclude this presentation of the work of the technological competition current, I would like to return to the following sentence by Foray, cited above: 'In the increasing returns to adoption regime, it is history (and not science or technology) that decides'. This economist's openness to history strikes me as highly promising. Yet a statement of this kind also has its limits. While it is necessary, as Aron (1957: 187) put it, to 'restore to the past the uncertainty of the future', it is also necessary constantly to examine the alternatives. Unlike the prevailing view, the history of technology is not a continuous and certain flow, but a series of changes and uncertainties. This is a perspective at the heart of the reasoning of our new economists. Their 'path-dependent choice' approach seems to be new and important.[8] It enables us to understand how a frame of reference is set and how it is blocked and becomes a convention that organizes the different actors' technological activity.

There is nevertheless one point on which real hesitations can be perceived in the position of economists of technological change: the scope of the historical phenomena to study. Should we be content to observe a technology and its diffusion, and consider all other phenomena as random events? Or should we, by contrast, extend the investigation? The more economists do history, as David does, the more their centres of interest are diversified. They would like to take into account individual histories which, like Edison's, develop specific strategies that transcend the technological domain. It is, moreover, advisable to include more general parts of history, as Cowan did with the evolution of the Cold War and the organization of military research in the United States. In my perspective, that is, the analysis of sociotechnical frames, it is clear that the historical viewpoint has to be extended to longer periods and to a wide variety of phenomena: technological necessity, political demands, social representations and so on.

Finally, as regards the technological competition perspective, it is important to bear in mind that not all periods of history have the same potential, and that certain options made early on last long after the constraints justifying the choice have disappeared. On the other hand, we have to avoid a deterministic view of the path-dependency phenomenon which suggests that, in the development of a new technology, the first to arrive is sure to win. Many examples exist in which the second or third system finally became the standard, as in the case of electricity, cited above.

EVOLUTIONISM

As Donald McCloskey (1976: 434) noted in a review, '[economists,] bemused by revolutions in the substance and method of economics, neglected the reading of history in favour of macroeconomics, mathematics, and statistics'.[9] Despite this dominant trend in economics, many economists have maintained their interest in history. Independently of the technological competition current examined above, there is another current whose interest is more macroeconomic. This includes so-called evolutionary economists as well as neo-Schumpeterians and researchers who study long cycles of economic activity. All these actors put technology and history at the centre of their analyses.

Rosenberg, a leading US historian–economist, wrote: ' The central theme, on which I wish to elaborate, is that technological improvement not only enters the structure of the economy through the main entrance ... but that it also employs numerous and less visible side and rear entrances where its arrival is unobtrusive, unannounced, unobserved, and uncelebrated' (Rosenberg, 1982a: 56). In the case of the steamboat, for instance, this technology incorporated developments of the steam engine and those of steel during its evolution. The propeller's final shape was the result of a process of trial and error.

This view of cumulative technological progress is not a characteristic of technologies in the nineteenth century only; it also exists with regard to contemporary high-tech. As Kenneth Knight noted, 'most of the developments in general-purpose digital computers resulted from small, undetectable improvements, but when they were combined they produced the fantastic advances that have occurred since 1940' (Knight, 1967: 493, cited by Rosenberg, 1982a).

This question of technological progress at a micro level is also found in the work of US evolutionary economists. Richard Nelson and Sidney Winter, the leaders of this current, distinguished two types of action in a firm's technological research. First, invention is contingent and uncertain; second, once it has been produced it alters the firm's knowledge: 'Something will be learned about a class or "neighborhood" of technologies' (Nelson and Winter, 1982: 249). These learning phenomena take place differently, depending on the firm, which probably explains why technological progress does not have the same effect on all firms. Evolutionary economists have the ambition of building a theory that links an analysis of invention to a macro analysis of technological progress. They are interested in research procedures and decision-making rules within the firm: 'At the micro level [technical change] refers to the characteristics of the learning process and the properties of a sort of "evolutionary hand"'. Giovanni Dosi and Luigi Orsenigo pointed out that in a fast-changing environment the 'evolutionary hand' plays more or less the same part as the 'invisible hand of the market' is seen to do in classical economics: 'It selects and orders the diversity always generated by technological and institutional change. Moreover, it is more powerful because it is not entirely invisible, but is forged within visible (indeed often dominant) technologies and institutions: it not only selects *ex post,* it also teaches and guides *ex ante*' (Dosi and Orsenigo, 1988: 32). Contrary to Adam Smith's famous metaphor, the 'evolutionary hand' is not simply a general principle; it has a very precise function in technological activity. We have here the concept of a framework of reference structuring daily technological activity.

As noted above, the question of technological change is set not only in a stable framework of reference but also in the context of its origins. This other form of evolutionism is at the heart of the original and isolated philosophical research of Gilbert Simondon, reported in his book *Du Mode d'existence des objets techniques*. The technical object, he said, 'is not a particular thing, given *hic et nunc*, but that from which it is born' (Simondon, 1989: 20). By defining the technical object in terms of its genesis and evolution, Simondon opposes the classificatory approach presented by many anthropologists of technology. The technical being 'can be the object of adequate knowledge only if that knowledge grasps the temporal sense of its evolution' (ibid.: 20).

The primitive technical object is 'the physical translation of an intellectual system' (ibid.: 46). Initially its form is 'abstract'; each part is treated as an absolute. The first technical problems to solve are those of compatibility between these wholes. Gradually the engineer will bring these different functions together in a structural unit. This convergence has two causes. One is economic: reduction of costs by standardization of parts but above all an internal necessity related to 'purely technical requirements: the object must not be self-destructive; it has to be maintained in a stable function for as long as possible' (ibid.: 26). 'The technical object progresses by internal redistribution of the functions into compatible units, replacing chance or the antagonism of the primitive distribution. Specialization is done by synergy rather than function by function'. The result is a 'concrete' technical object 'that is no longer at war with itself and in which no side effect harms the functioning of the whole' (ibid.: 34).

The inventor's initial project is altered through this 'concretization' process: 'Each part of the concrete object is not only that which by essence corresponds to the accomplishment of a function intended by the producer, but a part of a system in which a multitude of forces are at work and produce effects independent of the productive intention' (ibid.: 35).

The concrete technical object thus acquires an internal strength and solidity that makes it difficult to question. If we import this type of analysis into my own model, we can say that this concretization phase is an essential element in the constitution of the framework of functioning. We thus move from an abstract frame to a concrete one. The former is an object of debate and controversy; the latter is locked. Simondon enables us to explain this technological lock-in phase considered above. It is neither pure chance nor the result of the action of an exceptional individual like Edison, but the fruit of a particular process: concretization.

To grasp all the richness of the genealogical approach promoted by historian economists or philosophers, we have to be fully aware that a new technical object is rarely isolated. It is usually part of a cluster of innovations. To examine the development of an innovation or its impact, we need to study the system in which it took place. Earlier in this chapter I mentioned the quarrel between Edison and George Westinghouse concerning electricity. The various US historical studies on this period clearly show that the two inventors reasoned in terms of a system. Edison, for instance, simultaneously developed an electric bulb, a generator, a transport network and an electricity meter. Electric lighting started taking shape only after these four components had been articulated (see Hughes, 1983).

CHANGE AND A SWITCH OF PARADIGMS

Technological change as presented by the evolutionary current seems to be a continuous, cumulative movement, a series of incremental innovations. If, on the other hand, we focus on other events, the history of technology is characterized by uncertainty, discontinuity and a chain of radical innovations. The evolutionary approach clearly has difficulty in accounting for the latter perspective. In an evolutionary perspective, the question of origins is largely overlooked. Although Simondon affirms that a technological tradition clearly has an origin ('an absolute beginning', he wrote) (Simondon, 1989: 41), he tells us little about those beginnings. Economist Dosi completes the evolutionary perspective by introducing the notion of a technological paradigm. He is explicitly inspired by Thomas Kuhn's theories on science, mentioned above. Kuhn distinguishes two states of science: normal science and scientific revolution. Within an established paradigm, the 'normal scientist' refines concepts and measures the correspondence between facts and theory. When anomalies that cannot be explained by the existing paradigm appear, we enter into a period of crisis. The crisis will be resolved by the appearance of a new paradigm that will eventually be accepted by the entire scientific community. Kuhn saw the evolution of science (normal science–crisis–new paradigm) as taking place relatively autonomously within the scientific community. A paradigm is, moreover, less a set of theories than a mode of structuring the scientific community (Kuhn, 1962).

In 1973, 10 years before Dosi, historian Edward Constant used this concept of a technological paradigm. In Kuhn's theory the process of changing the paradigm starts with the awareness of an anomaly: 'that is;

with recognition that nature has somehow violated the paradigm-induced expectations that govern normal science' (ibid.: 52). The anomaly is then analysed and the process ends when 'the paradigm theory has been adjusted so that the anomalous has become the expected' (ibid.: 53). In his reflection on technological paradigms, Constant distinguishes two sources of technological anomaly: 'functional-failure anomaly' and 'presumptive anomaly'. The former is an extension of Kuhn's theory in the technological domain. The latter, by contrast, stems from a gap between the development of science and that of technology: 'That presumptive anomaly is deduced from science ... That scientific deduction is the sole reason for the sole guide to new paradigm creation. No functional failure exists; an anomaly is presumed to exist' (Constant, 1973: 555).

Yet on its own the functional-failure anomaly does not generate a technological revolution. It is necessary for a new 'candidate paradigm' to demonstrate the functional failure relative to the conventional system. In technology, unlike science, it is the new candidate paradigm and not the anomaly that provokes the crisis.

Constant has a highly internalist conception of technology:

Technological revolution is defined here only in terms of a relevant community of practitioners and has no connotation of social or economic magnitude. (Ibid.: 554)

... only at the level of community conversion ... only after new performance and cost criteria are established by initial development and practice of the new paradigm, can economic factors perform their conventional determinant role. (Ibid.: 559)

In Dosi's approach, a paradigm defines a technological trajectory. This can be described in a multidimensional space composed of economic and technical variables, like a range of possible directions that the paradigm defines: 'Each paradigm entails a definition of the relevant problems that must be tackled, the tasks to be fulfilled, a pattern of inquiry, the material technology to be used, and the type of basic artifacts to be developed and improved' (Dosi and Orsenigo, 1988: 16).

Choice is a complex issue. *Ex ante* it is very difficult to compare the worth of different paradigms. The market does not provide us with the information needed to make this choice, or even with that required to reduce uncertainty. On the other hand, certain elements of technological costs can be evaluated, although it is only *ex post* that the fecundity of the paradigm and its capacity to respond to the market can be determined. The situation in which the economic agent has to choose between several paradigms is rare, for it is in the nature of a paradigm to be imposed on engineers and to eliminate alternatives: 'The efforts and the technological *imaginaire* of engineers and of the organizations they are in are focused in rather precise directions while they are, so to speak, "blind" with respect to other technical possibilities' (Dosi, 1982: 153).

In fact the historian (Constant) and the economist (Dosi) have a fairly similar view of technological paradigms. In my own analytical model, a paradigm can be considered to constitute a frame of reference (some economists refer to 'tacit knowledge') in which the actions not only of innovators but also of users of the technique are set.

This concept of a paradigm has been highly successful among analysts of innovation and has often been used in a dynamic analytical perspective. Economist Devendra Sahal talks of 'technological guideposts' and 'innovation avenues' (Sahal, 1985). Nelson and Winter use related concepts of 'technological regime' and 'natural trajectory' (Nelson and Winter, 1982: 258–9). As an example of a 'technological regime' they take aeronautics in the 1920s and 1930s, characterized by the choice of the monoplane with an internal frame, the use of light alloys and special steels, retractable landing gear and so on.

They illustrate the concept of a natural trajectory by means of two far more generous examples: the exploitation of latent economies of scale, and the growth of mechanization and then automation. Yet these examples relate to phenomena of different amplitudes. The one case clearly corresponds to a paradigm since the firm uses the possibilities offered by technology to reduce costs. The other case concerns long-term phenomena found in a whole series of technological paradigms.

The taxonomy of innovations drawn up by Christopher Freeman and Carlota Perez (Freeman and Perez, 1988: 45–7) helps to define the specific nature of the appearance of a new technological paradigm as opposed to a new technical object. These two authors distinguish four types of innovation: incremental innovation, radical innovation, change of technical system and technological revolution.

'Incremental innovation' consists in a continuous series of actions which articulate technical opportunities in the framework of defined trajectories, to users' propositions. We are thus in a predefined technological frame which orientates the work of innovators, who are in a tactical position. Moreover, the demand and the market play an essential part.

The characteristics of 'radical innovation' are profoundly different. These are discontinuous events which are not situated in a defined technological frame. The idea of rupture linked to the concept of a paradigm is clearly present. Finally, radical innovation is far more dependent on the initiatives of R&D than on the pressure of demand.

The articulation between radical and incremental innovations produces 'technical systems', such as radio and television, for instance. These are present in several economic sectors or may even spawn a new sector.

Certain changes to the technical system are far more important. Freeman calls them 'technological revolutions'. This type of revolution differs from other innovations in that it 'not only leads to the emergence of a new range of products and services but also has an impact on all the other sectors of the economy, by modifying the structure of costs as well as the conditions of production and distribution throughout the economic system' (Freeman, 1986: 96). Freeman gives two examples of these technological revolutions: the railway and electric energy. These two techniques are 'technico-economic paradigms'. By contrast, Freeman sees the nuclear sector not as a technological revolution but as a 'technical system' in so far as this new technology has spread no further than the military domain and the production of energy.

The different levels of Freeman and Perez's taxonomy are of course debatable. For my purposes I simply wish to note the twofold distinction between innovation and frame of reference, and between frame of reference and long-term trend. Innovations are always set in a frame of reference. Either that frame is dominant and well established, in which case it is an ordinary technical activity (incremental innovation); or it is not yet well established, in which case radical innovation will help to get it off the ground. It would naturally be pointless to try to list the innovations and frames of reference that link the former to the latter. Everything depends on the analyst's focus. If I study the electric telegraph, electricity will be my frame of reference; if, on the other hand, I study Morse's or Bréguet's receiver, the frame of reference will be the electric telegraph.

Frames of reference are themselves set in long-term trends, not because super frames of reference exist but rather because their stability depends on the time-scale selected. There may be breaks in the framework, or else long, more or less underground technological or social movements may pave the way for a particular change of framework. As we have seen, the analysis of innovation and technological lock-in requires the integration of far more long-term trends.

There is another classification that distinguishes continuity and rupture in technology–production, on the one hand, and in market–customers, on the other (Abernathy and Clark, 1985). By articulating these two criteria, we obtain four groups of innovation: architectural innovation, which corresponds to a radical break, both in technology and in the market; revolutionary innovation, in which case the change concerns the technology but not the market; market niche innovation, characterized by technological continuity and real market innovation; and, finally, regular innovation, with a twofold incremental evolution in technology and in the market. This typology of economists seems interesting in so far as it is not content simply to distinguish ordinary innovations from changes in frames of reference. It separates technological changes from changes in use, and thus confirms the idea that the frame of functioning and the frame of use can evolve separately.

Long Cycles

The study of 'long cycles' is another area of cooperation between economists and historians of technology. Freeman's 'technico-economic paradigms' are set in the Schumpeterian tradition which considers that economic crises cause 'storms of creative destruction' which allow new innovations to boost growth.

This perspective has also interested certain historians of technology who have examined the distribution of major technical innovations in time. Gerhard Mensch has shown that this distribution is not random and that innovation-rich periods follow on from sterile ones (Mensch, 1979, 1988). Mensch identified three maxima in the innovation curve during the nineteenth and twentieth centuries: in 1825, 1886 and 1935. These three waves of innovation lasted about 15 years, each around a peak year. In years that were poor in radical innovation, as was the 1953–73 period, technical progress was orientated essentially towards improvement innovations.

Mensch tried to correlate this curve of major technical innovations[10] with the curve of scientific inventions. The underlying idea was that innovations took place later (by n years) than the corresponding inventions. This hypothesis is invalidated by the analysis. We thus have confirmation of what we have already noted several times: there is no automatic link between science and technology. Sometimes a technique uses certain possibilities offered by a scientific theory. In others, it is developed independently of science, and engineers construct their own theories in order to be able to advance in their work. Mensch's explanation for these waves of technological innovation is related to economic crises. He notes that innovation peaks are situated several years after major economic crises.

With the increase in unemployment and the underutilization of capital, hostility and mistrust regarding untried and risky new ideas disappear because people believe that anything might improve the situation. The result is a wave of fundamental innovations. This releases the economy from the technological dead-end in which stagnation kept it and leads to the revival phase. Mensch also shows that in these periods of innovation, the gap between the discovery of a principle and its commercial development shrinks. His approach fits with the tradition of Nikolai Kondratiev and

Schumpeter. The Russian economist demonstrated the existence of long cycles of economic activity that lasted about 50 years, while Schumpeter explained the appearance of those cycles in terms of technological evolution. The possibilities of technology are exhausted at the end of a cycle. During the crisis, innovation subjects the economy to a creative destruction that allows new growth.

Mensch's theory has the serious drawback of underevaluating the duration of the diffusion of innovation. As Freeman wrote: 'radical innovations cannot constitute the basis of major economic expansion because it takes one or more decades before the diffusion of innovations has perceptible effects on investments and employment' (Freeman, 1986: 96).

Should we, for all that, reject the Schumpeterian hypothesis of a link between innovation and long cycles? As Cristiano de Oliviera Domingues (1986) notes, no direct relationship can be established between technological progress and behaviour of production. Technological novelties appear randomly. These elements often remain overlooked, victims of their novelty. They are foreign to the frame of coherence of the dominant technological system of the day. Since the effects of this technological system gradually run out of steam, new technologies seem more exploitable, especially if they converge and if phenomena of synergy are created. This is when a 'technological revolution' occurs.

Like Domingues, Freeman articulates two theoretical traditions, that of Schumpeter on the effects of innovation on long cycles and that of historians of technology like Gille who studied technological systems. Mensch's mistake was to work on technological innovation lists observed in isolation, and to want to date innovations with precision. He considered, for instance, that radio appeared in 1922 (the year in which the regular broadcasting of radio programmes started), when in fact the preceding years were not simply a prehistory since Guglielmo Marconi's invention found a professional use from the beginning of the century. An innovation can be studied only from a sequential point of view.[11] Domingues is far more sensitive to this duration of innovation. He considers that the second part of Kondratiev's cycles, the phase of decline (1920–45 for example), is the period in which the subsequent technological system is developed.

In reality, the basic technologies of a system take longer to develop. The first technical achievements in the field of electricity started to be made in the late 1870s. In 1879, Siemens first demonstrated the electric railway. Edison installed the first electric lighting system in 1881 in Berlin. This was the end of Kondratiev's second cycle (1830–85). It was only at the end of the fourth cycle (1945 onwards) that electricity was effectively generalized to domestic use.

Some authors like Richard Barras (1986) and Freeman consider that we are currently on the brink of a fifth Kondratiev cycle, with a new technological system based on information technology (microelectronics, informatics and digital telecommunication). The key technical branch of this new system started to develop at the beginning of the fourth Kondratiev cycle (late 1940s and the 1950s) and was prepared at the end of the third cycle. Barras distinguishes two innovation cycles in the field of electronics and computer technology: one in the capital goods sector (industry) and the other in the consumer goods sector (general public). He shows that there is a half-cycle difference between these two types of good.

The reflection on long cycles briefly presented above led the specialists of technological evolution to incorporate economic and, more broadly, social trends into their analyses. Freeman studied connections between new technological paradigms and 'socio-institutional climates': 'The widespread generalization of the new technological paradigms is possible only after social institutions have adapted to the requirements of the new technology' (Freeman, 1985: 607).

Dockès questions this 'dichotomy between innovative dynamism within the techno-economic paradigm and the socio-economic context that has to be adjusted so that the overall change takes place and becomes efficient'. His opinion is that 'if the social dimension is present, it is "next to" and not within the productive paradigm' (Dockès, 1990: 39). Many of these analyses of long cycles have been criticized by the authors cited at the beginning of this book, for their radical break between the technological and the social. Dockès proposes, by contrast, the concept of a 'socio-technical paradigm' that he defines as 'the dominant way of thinking production'.

Similar perspectives are found in the work of Jacques Perrin who has studied technico-organizational paradigms: 'Any technological object has two sides: the *technological*, which is more apparent and allows the techniques used in its construction to be explained, and the *organizational*, which is generally unrecognized and makes it possible to identify all the relations and connections required for its production (or consumption)' (Perrin, 1991: 161: original emphasis).

Further on he notes that 'the capacity that technological artifacts have to change faster may give the impression that it is technical objects, and especially machines, that impose their constraints on organizations. This is not so at all; it is the same paradigm ... that orientates both the development of technical objects and that of organizations' (ibid.: 168).

The question of the distinction between technology and society is a point that I considered at length in the first chapters of this book. Clearly, the introduction of time into the analysis has no reason to fundamentally alter my point of view in this respect. Yet as soon as we adopt a dynamic perspective, we note that the different series do not develop at the same pace. Before the point of technological lock-in of a new socio-technical frame, the frames of functioning and of use can be altered according to very different time-scales.

NOTES

1. To justify this random connection between disciplines, I shall simply cite the following sentence by Pierre Dockès and Bernard Rosier: 'The most important link between history and economics lies in reflection on innovation' (Dockès and Rosier, 1991: 197). As for the link with philosophy, we simply need to recall the close bonds between that discipline and the history of science.
2. In his non-theoretical writings, especially *Industry and Trade* (1919), Marshall draws on detailed observation of the industrial dynamics at the time and adopts a different approach. I am therefore referring here to the theoretician of partial equilibrium.
3. It is important to differentiate between 'learning by using' and 'learning by doing', presented below.
4. On the importance of network externalities in the development of the North American telephone, see Mueller (1997).
5. In 1962, Kenneth Arrow wrote an article describing the mechanism of learning by doing (Arrow, 1962).
6. The case of QWERTY and Dvorak keyboards has also been studied by sociologists of diffusion. Everett Rogers cites the example of the Dvorak system to show that a rational innovation cannot be imposed (Rogers, 1983, 9–10).
7. This idea of articulation between chance and necessity is also found in Leah Lievrouw's approach. She proposes to link up the two very different schools of diffusion of innovations, and social shaping of technology, through 'a dynamic relationship between determination and contingency' (Lievrouw, 2006: 258).
8. Although this notion of lock-in also appears in historians' work. Marc Bloch noted: 'Once they have been created, institutions take on a form of rigidity and, bound by all sorts of ties to the social complex as a whole, grow strong enough roots to ensure that they are not ripped up once their initial *raison d'être* has disappeared' (Bloch, 1937, 13).
9. Robert Boyer, who commented on that article, adds: 'Perhaps historians have been too modest and have failed to practise the gentle symbolic terrorism that makes economist mathematicians so charming!' (Boyer, 1991: 1416).
10. The choice of basic innovations is obviously a subject of controversy between historians. Mensch therefore drew his conclusions with reference to innovation lists drawn up by different historians.
11. On Freeman's critical analysis of Schumpeter and Mensch, see Chapters 2 and 3 of Freeman et al. (1982).

6. Technological *imaginaire*

If we agree that technological choices depend on the path taken (see Chapter 5, above), one question remains: what is the origin of the path? Where should the investigations be aimed: at the first technological objects marketed? Or at the prototypes built by researchers, the intermediate mock-ups, the written projects, or even the utopias and dreams which feature as prominently on laboratory shelves as tools and fruitless prototypes? Utopias and dreams are not only a peculiarity of inventors; they concern far larger social groups who develop differing representations of the same technology. At the roots of a socio-technical context we find a whole series of imagined technological possibilities which seem to warrant investigation, not as the initial matrix of a new technology, but rather as one of the resources mobilized by the actors to construct a frame of reference.

GENEALOGICAL ILLUSION

'Objects are the incarnation of dreams' wrote Thierry Gaudin,

> Modern technology realizes age-old dreams: flight, mentioned by witches and wise men throughout the ages; ubiquity, through the proliferation of sound and images; even the Apocalypse, to which humanity has never been as close as it is now. All of man's great dreams, whether exhilarating or ghastly, are materialized in modern technology. (Gaudin, 1984: 13)

Although the relationship with the imaginary is a key element in the gestation of technological objects, there is little evidence that the contemporary technological *imaginaire* (social imagination) is, as Gaudin suggests, inscribed in the continuity of that of the Ancients. We cannot consider the first dreams of Santos Dumont as being similar to the myth of Icarus. For the one flying meant escaping, overlooking, being equal to the gods, while for the other it meant travelling through the air.

Jacques Perriault, in his genealogy of the mechanisms of visual illusion, went all the way back to Antiquity and notably associated the play of mirrors with magic lanterns. He found in each of these visual procedures 'the constant manifestation of the search for an illusion through ever more perfect simulations' (Perriault, 1989: 25). In compiling his genealogy he defined a precise rule: only select those inventors cited by their successors.

In the information field, Philippe Breton devoted himself to a similar task. He chose an 'upstream point of view', which meant focusing on 'those phenomena situated upstream from technological invention and which condition not only the dynamics of the invention itself, but also the conditions of its subsequent success – or failure – and, more broadly, the social meaning which the invention assumes in a given society' (Breton, 1992a: 45). Breton illustrates his thesis with the relationship between Norbert Wiener, one of the founders of computing, and the myth of the Golem (an artificial being imagined in Prague in the Middle Ages). For him the entire history of computing has been in keeping with the initial project: the construction of an artificial brain.

Thus, the genealogical approach places the emphasis on historical constants. It strives to reconstitute the matrix which spawned a technological object, considering that this matrix largely determines subsequent developments. Yet this approach, which introduces history into the study of technology and which is most appealing a priori, can be chosen only after having clarified an essential point: the influence of one inventor over another or the impact of a founding project on a technological tradition.

The notion of influence is not only used by the historians of technology, it also plays an essential role in much research work. The British art historian Michael Baxandall engaged in interesting reflection on this question, which may also prove particularly useful to the study of technology. '"Influence" is a curse of art criticism', he wrote, 'It is very strange that a term which such an incongruous astral background has come to play such a role' (Baxandall, 1985: 58–9). He sums up his reasoning as follows: 'If one says that X influenced Y it does seem that one is saying that X did something to Y rather than Y did something to X. But in the consideration of good pictures and painters the second is always the more lively reality' (ibid.). Behind the notion of influence there is that of causality. If one says that X influenced Y, one implies that X is the cause of Y. But Baxandall points out that in the action of the painter – and, we might add, of the engineer – the situation is exactly the opposite: Y is at the origin of the action and chose to use X.

By reversing X and Y, we not only put the phenomenon of influence back on its feet, but are also able to integrate multiple nuances. The link between Y and X may be expressed by 'drawn on, resort to, misunderstand' and so on (ibid). There are situations of workshops or learning where X triumphs over all the resources available to Y. In that case it is not X who is the cause of Y, but the organization of the artistic activity of the period, in the form of the studio, or the preponderant weight of academism. To illustrate his argument, Baxandall shows that it was not Paul Cézanne who influenced Pablo Picasso, but that Cézanne was part of the pictorial problems that the painter of the *Demoiselles d'Avignon* had chosen to confront, even if he selected only a few specific aspects of the work of the master of the *Holy Victory*.

Baxandall's analysis transcends the realm of art since he also applies it to the construction of a bridge (see Chapter 4, above). However, whether it concerns paintings or bridges, they are always original and specific objects. The technological activity discussed in this book refers, by contrast, to reproducible objects which, from one model to the next, all have very much the same components. My critique of the genealogical model or the technological influence model will therefore be slightly different. By examining the founding texts of a technology, the contemporary analyst often tends to relate technological projects or prototypes to today's objects and concepts. In a situation where these objects did not exist, it is totally unjustified to suggest that yesterday's research is an ancestor of today's objects. Claiming, for example, that the magic lantern is the ancestor of the cinema is probably enlightening for one of our contemporaries who knows nothing about the magic lantern, but it is of no help whatsoever in understanding the birth of the cinema. The whole dispute between the Lumière brothers (and other inventors) and Edison clearly shows that the cinema could well have developed along a path different from that of projection. The genealogical approach contains the serious risk of implying that history could not possibly have taken place differently because everything was already inscribed in an original matrix. This falsely deterministic point of view deprives of all its richness a history where at any moment the unforeseen could occur since inventors might well have chosen one solution rather than another.

Historians of technology sometimes note that a particular inventor did not understand the device he was creating. Instead of being surprised by what is an anomaly *a posteriori* only, one should start with the discrepancies between principles and technological creations in order to understand how a new object was gradually developed. Technological history is not always as clear and linear as the 'upstream approach' would have us believe. Beside the inventors whom official history was to baptize the 'founding fathers' – because, from today's point of view, their writings corresponded closest to what the technology they invented had become – there exist numerous lesser-known inventors whose technological doings were at least as important. The discourse of these inventors was stranger, attesting to the hesitations and debates which at the time attended the birth of a new technology.

Breton, in the context of his historical work on information processing, gathered together numerous writings by Wiener, Alan Turing and John von

Neumann, which he considered to be the founding texts on the computer: 'The modern computer is the product of a system of representations of the world and the values it spawns. The technical choices made ... far from adhering to a strict rationality inherent in technology, are determined by a system of thought' (Breton, 1993: 48; see also Breton, 1992b). In this case he is referring to cybernetics. The project was to build a machine which behaved like the human brain. In contrast with Breton's theses, the American historian Paul Ceruzzi considered that the inventors of the computer were incapable of foreseeing the multiple applications of their future machines. For them, the aim was rather to build powerful calculators intended solely for scientific and military purposes; about 10 such machines would adequately satisfy America's needs.

The idea of a universal machine capable of solving problems of a very different nature thus seemed to have little in common with the views of a computing pioneer such as Howard Aiken who wrote in 1956:

> [I]f it should ever turn out that the basic logics of a machine designed for the numerical solution of differential equations coincide with the logics of a machine intended to make bills for a department store, I would regard this as the most amazing coincidence that I have ever encountered. (Quoted by Ceruzzi, 1987: 197; see also Cerruzi 1983).

Here is a founding father who, unlike Wiener, did not imagine the computer as a universal machine.

THE SLOW BIRTH OF FRAMES OF REFERENCE

If we definitively jettison the idea of a genealogy of inventors, in terms of which a technology is said to have several brilliant founding fathers who, from the outset, had an intuitive understanding of its future characteristics, then we have to integrate history into the analysis in a different way. If I temporarily put aside the question of individual technological action, which broadly corresponds to the foregoing analysis by Baxandall, the way in which the frames of reference are constituted will provide me with my first element of reflection.

English historian and media sociologist Raymond Williams notes that most communicating machines begin their existence in a rudimentary technological form which will not survive (Williams, 1974: 19). With John Logie Baird's electromechanical television, for example, the frame of use of television was there; by contrast, the frame of functioning did not last and was replaced by a different one (the electronic tube was substituted for Baird's mechanical system[1]). While in some instances the frame of functioning evolved, in others it was the frame of use that changed.

Sometimes this frame of use is so different from ours that it has been completely forgotten. In a study of the first uses of electricity and, more specifically, of electric lighting, Carolyn Marvin writes: 'Assuming that the story could only have concluded with ourselves, we have banished from collective memory the variety of options a previous age saw spread before it in the pursuit of its fondest dreams' (Marvin, 1988: 154).

Lighting was certainly the application of the 'electricity fairy' that impressed people most in the nineteenth century. Like other innovations, electric light was first used in the public domain before spreading to the private sphere. In the 1880s this new form of lighting started to appear in the streets, and the illumination of shop windows was particularly successful. Theatres and auditoriums were the first to use it, before the electric show moved outside. We know that electricity played an important role in the universal shows and that exhibitions were devoted specifically to it (Paris, 1991; Philadelphia, 1994). Among the numerous wonders that contemporaries were able to discover were towers of light, illuminated fountains and gardens lit up night and day. Numerous events in public life also led to the building of electrical triumphal arches or the illumination of bridges. That of the Brooklyn bridge in New York was 'such a beautiful show, in the theatrical sense, that one of the most dynamic ferry companies in the city organized excursions by night called: "Theatre of New York Harbor by Electric Light"'.[2]

Electric light also served advertising. Signs composed of hundreds of electric bulbs started to appear on buildings. In Paris there were even projections of light. In London, magic lanterns projected images on Nelson's column – until this attack on the sacred aura of the glorious monument was prohibited. Numerous inventors imagined the projection of advertisements on the clouds and experiments were carried out in both Europe and the United States. Thus, electric light appeared as one of the first means of mass communication in public places. An event which clearly illustrates this frame of use of electricity was the celebration in New York of William McKinley's election as president in 1896. A huge crowd gathered in front of the *New York Times* building where the newspaper, connected to the national telegraph network, had direct access to the election results and projected a huge portrait of the new president.

Thus, at the end of the nineteenth century the frame of use of electricity was show business, the public sphere and, in a sense, a foreshadowing of mass communication. It was only a few decades later that a new frame of use was to appear: the private sphere, domestic comfort and home automation.

THE SOCIAL *IMAGINAIRE* OF TECHNOLOGY

Whereas the frame of functioning of a technology is developed primarily within the technological community and in research laboratories, the construction of the frame of use involves more diverse actors and is manifested in more varied discourse not only by technicians but also by 'literary persons': novelists, popularizers, journalists and so on. All this discourse contributes to the formulation of a social *imaginaire*. It is one of the basic components of the frame of use of a new technology.

The historian studies the social *imaginaire* of technology through a corpus of utopian discourse, science fiction novels and forecasts. Before studying the role which this literature may play in the construction of frames of use, we need to define some rules of method.

'Utopias' wrote Alphonse de Lamartine, 'are often no more than premature truths'. This romantic view of utopian texts corresponds, in a sense, to the 'upstream' approach considered above. Bronislaw Baczko, who studied the utopia of the French Revolution, rightly points out that in this perspective 'the Utopian is, so to speak, faced with a complete future. All that remains is to establish whether he has managed to "decipher" it or not' (Baczko, 1978: 16). This approach has another drawback: by focusing on the predictive function of utopia, we end up thinking that the effect of utopia stems from its realism. One should, on the contrary, consider that:

> [U]topias manifest and express, in a specific way, a given period, its fears and revolts, the scope of its expectations and the paths of the social *imaginaire* together with its way of envisaging the possible and the impossible. Surpassing social reality, even if in a dream and to escape from it, is part of that reality and provides a revealing testimony of it. (Ibid.: 18)

In any study of the social *imaginaire*, as in an analysis of technological choices, it is therefore essential to recognize in the past its entire richness and full potential, and to avoid explaining it and, *a fortiori*, judging it, in relation to future events.

Baczko was interested in the role of the social *imaginaire* in political change. However, to a large extent his analyses can be adapted to the case of technological change examined here: 'Even if social actors do not act out scenarios which they themselves imagined, their actions are nonetheless inseparable from their own images of themselves and their rivals, their dreams and their myths, their hopes and their fears' (Baczko, 1978: 404). This corresponds to Gaudin's approach in *Pouvoir du rêve*.[3]

The success of key innovations, like that of revolutions, mobilizes an entire society:

Setting the masses in motion, wrenching them from 'normal' life and projecting them from immobile history towards an accelerated history, cannot be done without the production of grand social mobilizing dreams and the symbols in which they are embodied, nor without the amplification of the tasks to be accomplished and the goals to be attained. Neither ideas nor dreams make revolutions, but how could they be made without the dreams they spawn? (Ibid: 404)

Political revolutions aside, we could say that the social *imaginaire* which constitutes the frame of use of a technology is common to innovators and users alike. It is, at a given time, that of a society or at least a part of it.

The Producers of an *Imaginaire*

In the nineteenth century, like today, an abundant literature was devoted to the imagined uses of new technologies. Utopian literature in the seventeenth and eighteenth centuries corresponded to the same fanciful artifice: the movement of the hero to another context where a radically different society existed. In the nineteenth century a new literary mechanism appeared: movement through time.[4] In these times utopian or 'Uchronia', technology often occupied an important place. Howard Segal studied American literary works that can be considered as technological utopias. He selected several texts which defined an ideal society where technology was a stairway to wealth and happiness. In the main, the technologies mentioned in these utopias did exist at the time of writing, but in the described contexts they were widespread and therefore at the centre of ideal societies. For Segal these technological utopian authors profoundly reflected the opinions and values at the heart of American society. Furthermore, he was struck by the coherence between the different utopias. 'Except for minor details, their separate visions are fundamentally alike. Their visions can be treated as one vision of America's future' (Segal, 1985: 32).

From 1883 to 1933, Segal identified 35 authors. These can be divided into two virtually identical groups: the technicians (inventors, engineers and supervisors) and the writers (journalists, novelists and teachers). In the group of technicians we find several personalities: two presidents of important scientific societies, a great inventor, King Camp Gillette, and other entrepreneurs of lesser renown. Among the writers only Edward Bellamy, with *Looking Backward* (1898), had achieved true literary success. The other inventors had written various technical books.

In Europe and especially in France, technical *imaginaire* literature appeared essentially in another form, that of science fiction novels. Jules Verne is obviously the best representative of this literary current. His works, like the utopian novels, were based on movement through time or space. Verne wrote about technologies which existed during his time and which he had already used personally – he had travelled, for example, on the *Great Eastern*, the largest ship at the time – and about others on which contemporary inventors were working. His interest in technology led him and a photographer, Nadar (pseudonym of Gaspard-Félix Tournachon), to found a 'Company for air navigation research'. While Nadar made several balloon ascents, Verne was content to make his imaginary characters fly. The writer, who bases his work on detailed studies, never hesitated to give his readers technical explanations. For example, in *Le Château des Carpathes* (1888) he wrote: 'The time has come to give an explanation of certain phenomena which have occurred during this account and the origin of which should soon be revealed. During this period [note that this story took place in one of the last years of the nineteenth century] the use of electricity, rightly considered to be "the soul of the universe", was perfected' (Verne, 1888 [1994]: 213). In *Twenty Thousand Leagues under the Sea* (1869), Verne had already entitled a chapter 'Everything by electricity'. This taste for the popularization of technology was not peculiar to Verne's work. In *Les Misérables* (1862), Victor Hugo wrote a whole chapter on the history of the construction of sewers, to introduce the peregrination of Jean Valjean in underground Paris.

Another figure who made his mark on science fiction was Albert Robida. The first novel by this caricaturist was a parody of Verne's *Extraordinary Journeys* (1878). He subsequently wrote two significant science fiction novels, *Le Vingtième Siècle* (1882), and *La Vie électrique* (1892). In no way can Robida be compared to Verne. Whereas the latter gave what were sometimes laborious technical justifications, the former took obvious pleasure in inventing imaginary solutions. It was less the technology which interested him than its uses (Lacaze, 1979: 77–89). His telephonoscope, for example, was an instrument for romantic conversations and for shopping, following the stock exchange rates or listening to the phonographic news bulletin at mealtimes.

This taste for technological utopias is not a peculiarity of science fiction specialists alone; it is also found in many late nineteenth century novelists. Villiers de L'Isle-Adam, in his *Ève future* first published in 1880, chose as his main hero one of his contemporaries: Edison. Whereas in 1817 Mary Shelley called upon a purely imaginary scientist, Frankenstein, to construct an artificial human being, Villiers de L'Isle-Adam used Edison to carry out the same task. He does of course advise the reader:

> Everyone knows today that an illustrious American inventor, Mr. Edison, has discovered over the past fifteen years or so a large number of things as strange as they are ingenious; among them are the Telephone, the Phonograph, the Microphone, and those admirable electric lights spread across the face of the earth, not to mention about a hundred other wonders. ... The (perfectly natural)

enthusiasm in his country and elsewhere has bestowed on him a sort of mysterious privilege, or very nearly so, in many minds. Does the CHARACTER in this legend – even during the lifetime of the man who inspired it – not therefore belong to human literature? (Villiers de L'Isle-Adam, 1986: 765)

Ève future is a philosophical and mystical fable on the way in which man can touch the mysteries of creation. The technological utopia of Émile Zola and Anatole France belong to a different perspective, that of a social, even socialist, vision of technological progress. In *Le Travail* (1901), Zola described new workshops which, thanks to electricity, were lighter, cleaner and quieter than before and which made profound changes in human work possible in the mid-twentieth century. In *Sur la pierre blanche* (1905), Anatole France, a man of letters and a book lover with no specific ties to technology, described a twenty-third century imaginary world in which 'the poets and novelists are published phonographically and [where] for the publication of theatrical shows a highly ingenious combination of the phono and the cinemato has been thought up, which reproduces the play and the actors' voices all together' (France, 1905 [1991]: 1123). Communication between individuals was also profoundly transformed: 'Telegraphy and wireless telephony were then in use from one end of Europe to the other and were so easy to use that even the poorest man could talk, when and as he wanted to, to another man at another point on the globe' (ibid.: 1114).

This taste for writing science fiction was also manifested in the work of one of the fathers of sociology, Gabriel Tarde. Émile Durkheim's main rival published *Fragments d'histoire future* in 1896.[5] In this utopia, which takes place in an underground world, men use technological resources not to increase the production of consumer goods, but to devote themselves to artistic activities.

These varied utopian fictions found among very different authors at the end of the nineteenth century may seem to be the result of a purely literary imagination. In fact this is hardly so, for inventors also dreamed of imaginary technologies and the use of their inventions. Two men of letters, who also had a truly technical activity, show through their own lives that a literary and a technological *imaginaire* can coincide more easily than we think.

Charles Cros, in the last third of the nineteenth century, was interested in communication devices. Most of his energy was devoted to colour photography, a field in which he obtained concrete results. He also wrote a report describing the principle of the phonograph but, unlike Edison, never built a prototype. His thesis on communication with the planets, registered in 1869 at the Science Academy (Cros, 1970: 510) is probably less known but more interesting from our point of view. Interplanetary communication was unquestionably an important issue for Cros. He imagined a system similar to that of headlights with a powerful electrical focus and lenses. His interest was also shown in the poems he wrote on the subject. Cros was not the only man of letters to be fascinated by this subject. According to Anatole France, with regard to intellectual discussions during that period, 'at the time we were ardent Darwinists ... The planet Mars also interested us and we were extremely curious about the conditions of life on it. We did not doubt that it was inhabited ... We were persuaded that one day we would establish contact with the men and primates of Mars' (France, quoted in Cros, 1970: 1221).

One might say that all this was merely literary musing, and that all the debate about contact with the planets finally proves that Cros was more of a poet than an engineer. Such a critique might perhaps be justified if the astronomer Camille Flammarion had not been equally interested in the same question.[6] Thirty years later, when wireless telephony had just appeared, William Preece, chief engineer of British Telegraphs, wrote: 'If a planet is inhabited by beings identical to ourselves and if they possess significant reserves of electrical energy utilizable in telegraphy, then it may be possible for us to have telephone conversations with the inhabitants of Mars' (Preece, 1898: 715, cited by Douglas in Corn, 1987: 35).[7] In the spring of 1919 Marconi announced that some of his radio stations had received very powerful signals 'seeming to come from beyond the earth'. The great inventor Nicolas Tesla was persuaded that they came from Mars (Douglas, ibid.: 54).

Hugo Gernsback, often considered to be the first American science fiction author, with his *Ralph 124C 41 +*[8] published in 1911, was also a technician and publisher. In 1906 he commercialized one of the first cheap wireless sets and, two years later, launched one of the earliest technical magazines on the radio. It was in this magazine that he published his novel as a serial. The description of futuristic technologies featured prominently in it with, in particular, a precise description of the principles of what, thirty years later, was to become radar. Twenty years later Gernsback launched the first magazine specialized in science fiction and for two decades he was to remain one of the main American publishers in this field.

The examples of the American technological utopians, French science fiction novelists, and Cros and Gernsback, show that at the end of the nineteenth century men of letters and engineers alike contributed towards the definition of a relatively unified technological *imaginaire*. This was apparent not only in the production of utopias but also, more commonly, in the futuristic discourse which no longer used fiction as a mediator for talking about the future. At the beginning of the century Amos Dolbear, one of the inventors of the telephone, wrote a text entitled 'Electricity and civilization' in which he stated that 'each system which broadens the environment of everyone and makes the rest of the world his neighbor, fulfils a true civilizing mission' (quoted by Marvin, 1988: 192). Edison similarly made numerous declarations on the future uses of the technologies he invented.

Apart from inventors and novelists, there existed a third source for this technological *imaginaire*: popular magazines and, in general, the entire press. The latter did not limit itself to presenting inventions and imagining the uses indicated by their authors; it also participated in the construction of a technological *imaginaire*, for example by circulating rumours about inventions which did not even exist. Thus, in 1877, a year after Bell's invention of the telephone, the famous French popularizer Louis Figuier announced that Bell had developed the 'telectroscope' for communication by means of sound and image. A year later *Punch* reported that Edison had invented a similar device (quoted by Abramson, 1987: 7–8). Until the end of the century many such announcements were made, associated with different inventors. The press also played an important role in the construction of imaginary uses. It described at length the multiple uses of electricity in the future factory and the home. The particularly rich technological *imaginaire* of the late nineteenth century associated existing technological devices with others that were no more than dreams or plans, and real uses with fanciful ones, so that realistic forecasts were difficult to make: 'Even among observers with scientific expertise', notes the historian Carolyn Marvin, 'few in the volatile atmosphere of the late nineteenth century could be certain what wild fantasies might already have been translated into technical realities' (Marvin, 1988: 194).

The study of the social *imaginaire* of technology therefore appears to be an important component in the study of innovation. Several American historians have started to develop a programme along these lines. Since the seminal work of Leo Marx (1964) on the steam engine, research has been undertaken more recently by Marvin (1988) on electricity, Susan Douglas (1987) on the wireless, Joseph Corn (1983) on aviation and Rosalind Williams (1990) on the theme of the underground in the nineteenth century. As Corn wrote in the conclusion to a collective work on the technological *imaginaire*, the most naive fantasies 'are part of the same cultural milieu in which actual invention takes place and technology is adopted and diffused. Although scholars usually explain the activity of inventors as a response to market demand, it is plausible that inventors have responded as well to popular dreams and expectations' (Corn, 1987: 228).

Discourse of the Imagination

In the preceding section we saw who were the authors and supporters of discourse on the technological *imaginaire*. I now propose to study, as an example, the content of two of these discourses: the discourse attending the birth of the steam engine and that which surrounded the emergence of electricity.

Marx studied what he called the 'rhetoric of the technological sublime' (Marx, 1964: 195) which appeared in the mid-nineteenth century in the United States with the steam engine. The locomotive was associated with the idea of power and speed and pictures of the iron horse and the Titan of Fire were often used. One expression constantly recurs in the literature of the time to characterize the railway and the telegraph: the negation of space and time. Yet although these technologies were new, they were still in keeping with the perspective of the Enlightenment and the belief in progress. Inventors became the greatest heroes of the time and took the place of poets in the popular imagination. Finally, there was some sort of affinity between new technology and democratic functioning.[9] While new machines strengthened democracy, the citizens of a democratic country were also the main beneficiaries of the new technology. In short, as a foreign visitor noted, the railway appeared 'to be the personification of what was American' (Chevalier, 1836: 208).

The 'electricity *imaginaire*' was, to a fairly large extent, opposed by that related to the steam engine. Electricity, which was often at the centre of late nineteenth-century technological utopias, was associated with themes such as the dispersion of the population across the territory, or the diffusion of culture. Scottish writer Patrick Geddes, a biologist and urban planner, was also one of the first to present electricity as a technology which would spawn a new industrial era opposed to that of the steam engine. With the entry into the neo-technological era, man would be able to build a utopia.[10]

This point of view was shared by inventors such as Werner von Siemens, who wrote:

> Now is the time to build electric power stations throughout the world ... Thereby the small workshop and the individual working by himself in his own home will be in a position ... to compete with the factories that generate their power cheaply by steam engines and gas engines [and] will in the course of time produce a complete revolution in our conditions, favour small scale industry, add to the amenities and ease of life' (cited by Carey and Quirk, 1970: 230).

This theme was adopted extensively by the media. Marvin, who made a detailed study of the press at the time, notes as one of the main themes in public imagination around electricity, that it was expected to: 'reverse the centralization of production in factories, lead to the rise of clean cottage industries, unify the home and the workplace, and lower the divorce rate' (Marvin, 1987: 203).

Politicians were also to take up the theme of a 'new electric society'. In the 1930s Stuart Chase, a representative of the US League for Industrial Democracy, published an article entitled 'A vision in kilowatts': 'Electricity', he noted, 'can, on a universal scale, give us a high standard of living, new enjoyable occupations, entertainment, freedom and the end to tiresome work' *Fortune Magazine* (Chase, 1933, p. 5). Electricity thus became a theme of political rhetoric. The same was true in Russia. We recall Lenin's famous statement which made the Soviets and electricity the two characteristics of socialism. In the United States, the New Deal also promoted electricity with the creation of the Tennessee Valley Authority and the Rural Electrification Administration. Theodore Roosevelt indicated in one of his speeches that electricity could lead to an industrial and social revolution (Carey and Quirk, 1970: 235).

Since this *imaginaire* concerning electricity was so widespread among utopians, engineers, journalists and politicians alike, why, ask the historians, did their forecasts turn out to be so wrong? For Thomas Hughes (1987), industrialists and urban authorities used electricity to strengthen the concentration that they had already started. Technology led to no social revolution; it adapted to the dominant trends in society. From this point of view the technological *imaginaire* might appear as pure reverie with no consequences; but that is not so. An approach in which one studies technological utopias only in so far as they are supposed to give exact forecasts for the future, is obviously of little use. The history of the *imaginaire*, like that of technology, is not unilinear. On the contrary, representations of technology are diverse and conflicting. Furthermore, links in the technological *imaginaire* are even more diverse and obscure than technical links.

In an interesting essay on the technological *imaginaire*, Rosalind Williams notes with regard to electrical utopias: 'The prophets of regional decentralization, like so many other social prophets of the late nineteenth century and early twentieth, assumed we would espouse the values of order, efficiency and rational planning. Instead, fantasy worlds lure us on every side' (Williams, 1990: 113). In short, technological modernity did not allow this nostalgic return to a rural society reconciled with itself. On the contrary, it built a new space, that of the city, a place of pleasure. Williams contrasts this urban utopia with decentralized electric utopias. A literary figure marvellously illustrates this urban mythology: Charles Baudelaire as analysed by Walter Benjamin (1989). Baudelaire's Paris is an allegorical city. The lighting (as at the time) organized it into an interior space in which the '*flâneur*' could stroll. Separations between the inside and the outside faded, the department store was a type of road and the road, with its arcades, an interior space.

These closed spaces were the world of artificial pleasures offered by industry. Williams notes:

> Despite these ambiguities the fantasy of the enclosed artificial environment has flourished, primarily because it is so marketable ... The pseudo-subterranean high-tech *salon* that Beckford, Bulwer-Lytton, Verne and Tarde imagined as a refuge from reality, can even more clearly be recognized in the first-class airplane cabin, the hotel suite, the limousine, the executive office, the fine restaurant, the shopping mall. (Williams, 1990: 113)

Thus, the technological *imaginaire* varied. For some, electricity was associated with a nostalgic return to a technologized rural sphere; for others it symbolized the urban sphere of luxury and consumption. Likewise, the steam engine symbolized power for some and hell for others.[11] Through these two examples we clearly see that the social representations of a single technology are often contradictory. Different technological *imaginaires* produce contrasting visions of the future. Reflection on relations between the technological *imaginaire* and innovation must therefore include the interaction between these different representations.

ROLE OF THE *IMAGINAIRE* IN THE PROCESS OF TECHNOLOGICAL DEVELOPMENT

Does the technological *imaginaire* play a part in the development of innovation? To answer this question we first need to take on board one of the results of the foregoing analyses, and assume that the technological *imaginaire*, in the form of different antagonistic components, is often shared by innovators and users alike. (This bears little relation to a deterministic schema where the founding fathers of a technology or manipulating industrialists diffuse their conceptions of the new technology throughout society.) Debate on the place of the technological *imaginaire* cannot then be limited – as is often the case – to a statement on the degree to which technological forecasts were realized. If we wish to determine the role of the *imaginaire*, we carefully need to study the representations of the designers and first users of a technical object, and to see how options in the design or use are influenced by these representations. The latter constitute as many elements as the different users will employ to promote a specific technological solution, to change it or to block a particular project. Under no circumstances does the technological *imaginaire* constitute the mere embryo of a future sociotechnical frame of reference. It is a resource available for the actors just as known physical phenomena or existing social practices are.

Some sociologists of science have reached the same conclusion. They believe that expectations of technical and social potential, that is to say, promises, constitute a rhetorical 'space' in which the relevant audience can be mobilized (van Lente and Rip, 1998).

If we look at the history of techniques, Claude Chappe, the inventor of the optical telegraph, was for example fully aware that the new representations of space and of the state which appeared during the French Revolution were a key element for defending his project. He wrote explicitly to Joseph Lakanal, his spokesman representing him in the parliament: 'How is it that [our opponents] have not been struck by the ingenious idea that you developed yesterday ...? The establishment of the telegraph is indeed the best answer to those publicists who think that France is too spread out to form a Republic' (Flichy, 1995: 9). Edison had the same perspective. The great American inventor and his associates were not only making electricity, 'they were also making meanings' (Bazerman, 1999: 333). This creation of meaning has to take place in various arenas: laboratory, patent system, investment bankers, journalists, families and so on. For Bazerman, who studies how Edison gave sense and value to electrical light and power, inventors must incorporate their technology into existing cultural systems of meaning. But, at the end of the process, meanings and values are permanently set. We could cite many more examples of how inventors mobilized different representations in the construction of a socio-technical frame.

Finally, we need to look at 'the ideological frameworks within which emerging technologies evolve' (Douglas, 1987: xvii). JoAnne Yates showed in a seminal book (Yates, 1989), that a new managerial ideology played a significant part in the development of information technologies (typewriter, duplicating machine, mecanographic machine) in the second half of the nineteenth century. This new ideology appeared as a response to the coordination crisis facing firms in which the number of employees rose too fast. Written procedures then constituted the most effective mode of coordination. In this respect, 'information machines' seemed necessary both for immediate use and for compiling reference statistics. Information technologies had become the symbol of modern methods and could be used even when they were not economically justified: 'It was then widely diffused among managers and firms by publications, associations, consultants and contacts, serving as a template for managerial strategy and practice. This ideology encouraged and was encouraged by information technology and technique in a mutually reinforcing dynamic' (Yates, 1994: 27).

While this work by Yates focuses mainly on the role of ideology in the diffusion of a technology, I would like to examine the role of the *imaginaire* in technological development itself. To that end, I have chosen other, more contemporary information technologies, the microcomputer and the

Internet, and focus specifically on their relationship with the American counter-culture.

The Personal Computer *Imaginaire*

My study of the first example draws upon the very thorough research of the journalist Steven Levy (Levy, 1985). Rather than focusing on the founders of Apple or Microsoft and writing new hagiographies, like many of his confrères, Levy devoted a book to hackers, those computer hobbyists who were particularly active on the US West Coast in the 1970s. Levy's account was organized not around inventors but around the places of interaction: clubs and magazines. That was where diverse projects and utopias intersected; they were one of the main loci for the construction of the microcomputing socio-technical frame.

In the American counter-cultural movements in the early 1970s, computing seemed to be related essentially to the army, to major corporations and, more broadly, to a centralized system of social control. 'Computers', we read in the publications of the time, 'are mostly used against people instead of for people, used to control people instead of to free them' (ibid.: 172). This led some independent computer specialists to imagine a radically different use: 'a communication system which allows people to make contact with each other on the basis of mutually expressed interests, without having to cede judgment to third parties' (ibid.: 156). In this way it was thought that information processing could be used as a guerrilla tool against all bureaucracy. This information utopia was to spawn several alternative bodies from 1971 to 1972. Resource One, for example, undertook to compile an urban alternatives database, a directory of social services and community groups. The People's Computer Company created a shop in which the inhabitants of the district could come to learn the rudiments of computer use. Finally, Community Memory (a new branch of Resource One) created another shop with a simplified terminal through which people could consult advertisements on a database and key in their own messages. All these alternative projects relied on existing equipment; they used their own minicomputers or machine time on mainframes. However, this type of system had several drawbacks. Computer tools such as these were particularly opaque in so far as, in the event of a programming hitch, the user was simply given an error message, without any further indication. Moreover, such machines could not be widely diffused throughout society.

It was therefore necessary to build a new type of computer. Lee Felsenstein, one of the leaders of Community Memory, decided in 1974 to embark on this adventure. His idea was to build a tool capable of adjusting to the specific requirements of users, in line with the principles formulated by Ivan

Illich of whom he was a faithful reader. In 1971 the development, by Intel engineers, of a new component, the microprocessor, afforded a suitable opportunity for these new computer builders. However, to achieve his decentralizing conception of computer technology, Felsenstein had in mind a complex architecture where several processors would work in cooperation. He was not the only one at the time to imagine a new breed of computers. Other young engineers working in computer companies spent their evenings and sometimes their nights on similar projects. Most of them spoke about it very little for the whole idea usually seemed crazy to reasonable people. The proof is that Intel had no intention whatsoever of having its microprocessor used in that way. At the beginning of 1975 a group was formed, the Homebrew Computer Club, of which Felsenstein soon became the leading light. The aim of this home computing club was to bring together those hackers who wanted to build a computer or a visual display unit (VDU). At the first meeting it emerged that six of the 32 participants had already built a system similar to a microcomputer. Suddenly a practice which had been individual and underground became social. The foundations of a frame of reference had been laid.

Two months earlier a New York magazine for hobbyists, *Popular Electronics*, published an article on the Altair, a computer which could be purchased in kit form from a small, unknown business in New Mexico. The company was targeting a market of 400 units. Orders topped this figure the first day, despite the Altair being very difficult to use (data had to be introduced individually in binary language and the result of the operation had to be deciphered from pilot lights flickering on and off). The Altair was therefore above all a machine for those who wanted to build their computer at home. It was to be the focus of debate at the Homebrew Computer Club where many discussions concerned peripheral elements to be added, for example, extra memory, a VDU and a Basic interpreter to facilitate programming.

The Homebrew Computer Club proved to be a particularly fertile place of debate. Not only did many participants build their own machines there, several of them also developed commercial machines: Felsenstein's Sol (1976) and Steve Wozniak's Apple II.

These hackers, who were profoundly attached to a culture of exchange, sharing and conviviality, reacted very differently when it came to 'launching a business'. Felsenstein rented the services of a fledgling enterprise and reinvested his income in a project to revive the Community Memory. Wozniak, who had originally built a computer for his own enjoyment, was reluctant to embark on the creation of Apple with his friend Steve Jobs.[12] In the software field the hackers' ethic of free circulation of information was confronted by the strategy of the new entrepreneurs. While some software designers sold a sample for a nominal sum to those who requested it (a few dollars for copy charges), others such as Bill Gates claimed royalties. The former attitude was nevertheless the most widespread, as Apple initially provided the Basic interpreter and its documentation upon demand.

This conflict between two value systems was obviously not peculiar to the pioneers of microcomputing; it was found among many representatives of the 1968 generation. Even more interesting is the way in which conflict was resolved. Whereas some used the microcomputer to revive their project of computer democracy, others became new captains of industry. From the counter-cultural tradition they nevertheless retained the idea that microcomputing had to be convivial and decentralized: two key elements in the new socio-technical frame. Apple, in particular, was the bearer of this image of microcomputing, not because it was fundamentally different from other companies, but because it was able to successfully build its image around this theme.

In 1977, with Apple, the microcomputer left the world of hobbyists to become a home computer. It was the start of the mass diffusion of this new machine and its frame of reference. The governor of California, Jerry Brown, who had a passion for the new technology, declared in 1978 that the microcomputer was part of 'an entirely different culture' and that 'information is an equalizer and destroys hierarchies' (Brown, 1978, cited in Roszack, 1986: 144).

Three conclusions can be drawn from this history of the microcomputer.

First, the path on which a stable technology depends is not linear as in the 'upstream' point of view or even tree-shaped as with Paul David; it consists of an abundance of possibilities. At the origin of microcomputing we find multiple projects running parallel, crossing over one another and sometimes even supporting one another. There were the hackers who wanted to invent popular computing, although some simply wanted to meet a technological challenge, and there were the professional researchers who wanted to exploit the microprocessor's potentialities. The engineers of the large firms never managed to convince their corporate managers, while the inventors–entrepreneurs produced the first microcomputers.

Second, innovation, in the sense of production of a frame of reference, is a fundamentally collective activity. At the origin of technology there are not just a few heroic, unknown and solitary inventors who fill pages of plans in their rooms at night or tinker in their garages over weekends. There are, above all, also places of socialization where these projects are presented and where, from the encounters between these different hypotheses, a first frame of reference is born. In the specific case of microcomputing, these places of socialization were essentially clubs and counter-cultural magazines. In other cases there were primarily legitimate technical journals, conferences, exhibitions, standardization institutions and so on, all of which participated in the production of frames of reference.

Third, the technological *imaginaire* is a key component of the development of technology. Without the myths produced by the American counter-culture in the early 1970s, the microcomputer would probably have remained a mere curiosity. The ideology of computing for all, in a decentralized form, suddenly lent a whole new dimension to Wozniak's tinkering in his garage. However, the influence of the counter-culture on the microcomputing project makes sense only if we adopt an approach similar to that of Baxandall. That is to say, it was Wozniak, Steve Jobs and many others who chose to associate the values of the counter-culture with their passion for computing. They defined the problems that they chose to confront. The community ideology alone did not create the microcomputer; at best, it produced a mythical frame of use. In Levy's account, one person clearly expressed this weakness of the technological dream: Fred Moore, one of the founders of the Homebrew Computer Club, for whom this group was essentially a place of conviviality, for exchange on a non-market basis. Tinkering on computers hardly interested him, and he ended up leaving the club. The basis of the hackers' activity was immersion both in the counter-culture and in the world of computer tinkering; these two components were not only juxtaposed but also very closely linked. The ties needed for the establishment of a permanent socio-technical frame were built by actors; the technological or social dreams had no power other than supplying the resources for the action.

Internet *Imaginaire*

Whereas individuals and very small groups were at the origin of personal computing, the early Internet was characterized by a larger project launched by academics funded primarily by the Advanced Research Projects Agency (ARPA) the fundamental research agency of the US Department of Defense (Abbate, 1999). The project was initially defined by the first managers of ARPA's computer department, Joseph Licklider and Robert Taylor, both from leading universities in this field. They then mobilized colleagues around this operation. Most of them believed that human–machine symbiosis had to be created and that computers could be connected to other machines in a network organization. Many of their statements made at the time contrasted the contemporary age of computing with a new period in which humans would be able to have a dialogue with machines and cooperate with their peers via information networks.

As the first designers of Arpanet (the network financed by ARPA) started to achieve that initial goal, by fleshing out their project, they also defined it with more precision, altered it and reorientated it. Gradually, as the network developed, a common *imaginaire* was constructed.

The principle of cooperation underlying the network architecture was also a golden rule for the functioning of the project. The cooperative *imaginaire* of the Internet was particularly strong from the outset since it applied both to its technical aspects and to its social organization. For instance, the Network Working Group in charge of defining communication protocols headed the minutes of its meetings 'Requests for Comments' (RFC). These soon became the base of the technical documentation and standardization of the Internet.

Reflecting on the mode of cooperation that they had established between themselves, the designers of the Internet naturally raised questions on uses of the network. Here again, their experience in the construction of this new information system led them to add substance to their projects and to alter them. Whereas Arpanet was initially intended primarily for distant access to the data-processing capability provided by other computers, it gradually came to be seen as a tool for interaction and cooperation. Licklider introduced the notion of 'online community', consisting 'not of common locations but of common interest' (Licklider and Taylor, 1968 [1990]: 21). This community of computer experts is a very particular example of a technical project in which the same actors endeavour to conceive of, build and use a technical device. The *imaginaire* of a cooperative information technology is both the initial utopia that enables the project to be launched, and the result of an imaginary collective construction based on technical creations and experimental use.

This imagined cooperation appears not only in the Arpanet community but also in other groups of computer specialists. Academics cooperating regularly with management and organization specialists developed a computerized conferencing system. Nine years after the launching of Arpanet, two of them, Starr Roxane Hiltz and Murray Turoff (1978), published *The Network Nation: Human Communication via Computer.* Through a series of examples, this book showed how a human group could use computerized conferencing to exercise 'collective intelligence'. In the world of the Californian counter-culture, computerized conferencing took on a new form as it was tested as a means of communication open to all, outside professional contexts. In this respect, Howard Rheingold, one of the leaders of a well-known project, The Well, spoke of a 'virtual community'. He used the same argument as Licklider on communities of interest, but took it further. Virtual communities, he maintained, were superior to traditional communities in so far as they enabled people to directly find others with the same values and interests.

During the same period, various computerization movements were born (Kling, 1996), and hackers developed a program enabling microcomputers to communicate via the telephone network. This was the beginning of BBS (bulletin board system), linked up in the 1980s via a cooperative network. The founder of this network claimed that its design was 'explicitly based upon anarchist social principles' (Tom Jennings, 1998) and indeed, BBSs were not only information sources but also nodes in the network, which meant that they could circulate their own messages and even form their own networks.

Whereas Rheingold and the founder of the BBS cooperative network saw the Net as a means to create virtual communities without any geographic anchorage, consisting of very distant individuals, leaders of communities based in urban or rural areas saw network computing as a tool to revive their local identity. As one of them said: 'BBSs help integrate a community'. Likewise, the Santa Monica PEN network proposed 'to enhance the sense of the community' in the city (Frank Odasz, 1995).

These different forms of community and cooperative *imaginaire* that emerged through these experiences belonged not only to the designers and organizers of these BBSs, but also to users. A user of The Well wondered: 'Can people come to have emotional attachments to one another without ever facing each other?'. 'The answer is an emphatic yes', as in virtual communities. A user of a BBS community commented that the community had 'put a human face on an otherwise impersonal city. I now feel that my home is directly connected to thousands of friends through my modem' (Flichy, 2007: 74–81).

Despite these differences, during the 1970s and 1980s the Internet *imaginaire* was forged by project leaders as well as computer scientists. This *imaginaire* evolved as the network took shape. Notwithstanding the technical and organizational specificity of each of the projects mentioned here, the different actors' representations were fairly similar.

This coherence of representations is partially explained by the fact that the different computer network projects developed in relatively homogeneous social worlds: university, counter-culture, community movements and so on. But what happened when, in the 1990s, the Internet became a mass product with widely diverse users? A new discourse on network computing and its social impact appeared. It was not the discourse of computer scientists or the first users, but that of specialists in words, experts and journalists writing in the computer press or in publications for the general public.

Contrary to certain beliefs, the main function of this new discourse was neither promotional nor simply to accompany the diffusion of the technology; it was rather a reformulation of the *imaginaire* of the preceding phase. The digital intelligentsia that produced it were already seasoned users who were thoroughly familiar with the Internet; they were not going to build utopias cut off from technological reality. These intellectuals were to serve as mediators between designers and users, and to participate in the definition of the new socio-technical frame and the stabilization of the new medium.

The Virtual Community (Rheingold, 1994) is a good example of this new discourse. It was the first book on the Internet that was neither a technical volume nor a handbook. In it the author wrote at length about The Well and his own experience as a user and newsgroup leader. He also presented other electronic communities and Arpanet. Through this account he constructed a representation of the Net that combined characteristics of these different experiences. Virtual communities brought together individuals from the four corners of the globe, although most of them remained locally embedded. They developed conversations that were as intellectually and emotionally rich as those in real life. This, Rheingold maintained, was a balanced form of interaction between equals. In the final analysis, the NET can serve to recreate the disintegrating social link, to breathe new life into the public debate and, more generally, to revive democratic life.

Rheingold's book thus proposed one of the founding myths of the Internet. Rheingold turned a few experiences (Arpanet, The Well) into the Internet reference model, but failed to see that its expansion to a much broader social space profoundly altered the situation: the mode of functioning of counter-cultural communities and academic circles was obviously not the same as that of an entire society. This period was characterized by a rapid evolution of Internet *imaginaires*. Computer-mediated communication was no longer experimentation with a technical project, aimed at mobilizing a small group of academics; it offered American society new relations of communication which, until then, had been experienced only in specific groups.

In parallel with Rheingold's book, computer scientists who were thoroughly familiar with the new medium wrote various introductory handbooks for Internet users. From autumn 1993 the mainstream media started to present the Internet as a means of communication for the public at large. These different books generally suggested that the mode of electronic sociability that had developed in the university environment and BBSs could spread to the ordinary world. Although this discourse was filled with illusions, it did have the advantage of presenting possible uses of the Net. It was particularly credible since such practices actually existed and were known to those who had access to the network.

The Place of Internet *Imaginaire* in Design and Uses

Now that I have shown that there was coherence between these different Internet *imaginaires*, we shall turn to the relationship between these discourses and the technical practices of design and use. In fact, this *imaginaire* corresponded as much to pure fantasy describing a technically impossible world and fanciful ideas of use, as to precise projects for new devices and new uses. Even though sometimes illusory utopias[13] made it possible to break away from reality, it was through the description of uses and their experimentation that the new technology was stabilized.

To understand the role of these discourses in the conception and diffusion of the technology, we need to look at the people for whom it was intended. For which public did it cater? What was the size of that public? Some texts were clearly written for a few individuals, whereas others were published in journals or scientific collections with a limited circulation or else in the mainstream press. The addressees can usually be identified. They were either actors in the design process (colleagues, collaborators, directors, financiers and so on), or else potential users or a large public that simply wanted an opinion on the new technology. Finally, in certain cases the authors were hardly interested in their public; they wrote for themselves above all, to clarify their own ideas.

The different discourses on the Internet presented above corresponded to identified addressees. The writings of computer scientists were intended for their own community. Those of project leaders like Licklider or Taylor served to mobilize participants in the Arpanet project or to convince the ARPA management. Ordinary designers who wrote RFC launched a debate with their peers. All these texts defined successive iterations of a computer network project. They made it possible to set the socio-technical framework of this new kind of computer technology, the frame of functioning and the frame of use.

This discourse was particularly effective since computer programming, unlike other technical domains, required researchers to invest their time, above all. Researchers could therefore test their own hypotheses and see the results. Via the Internet, they could also transmit their programs to peers and ask them to test them. In the social world of academic computer scientists, as well as in that of hackers, legitimacy was most often acquired through the software that individuals produced rather than through the texts they wrote.[14] This was therefore a situation in which the gap between discourse and professional practices was small.

As soon as we leave the domain of technical creation and move over into that of use, we see that computer specialists' discourse was twofold: they described their own practices, since as designers of Arpanet they were the first users, and they imagined uses that extended beyond network computing to society as a whole, in the professional, private and public spheres alike.

Discourses intended for the general public also had several functions. Above all, they endeavoured to describe a particular social world, that of the computing counter-culture. They then sought to present it as a model for American society as a whole. Lastly, they tried to promote the new technology. The journal *Wired*, which was launched in 1993 when the Internet was spreading to the general public, and had a large circulation in the more fashionable middle classes, was instrumental in conveying the cyber-*imaginaire*. But *Wired* was not the only mediator of the cyber culture; similar discourses were also published, albeit slightly later, in the US news magazines. *Time* entitled the cover story of an issue in early 1993 'Cyberpunk'. The magazine interviewed former hippies converted to computer network technology, especially Rheingold who was presented as 'the first citizen of the Internet'. As for *Newsweek*, it declared 1995 Internet Year. Its end-of-year issue opened with the following sentence spread over four pages: 'This changes ... everything'.

But the news magazines were not content simply to disseminate the representations of Net designers; they also promoted its diffusion. *Time* commented, for instance, that 'suddenly, the Internet is the place to be', while *Newsweek* published an article entitled 'Making sense of the Internet' (Flichy, 2007: 92–5).

Discourses on the Internet *imaginaire* seemed thus to be profoundly articulated to the development and use of this new technology. Some of the designers' initial intentions materialized. By using this intellectual tool themselves, they showed what could be done with it. When the mediators recounted these early experiences, they presented a frame of interpretation and action for the general public. We can thus conclude that discourses on the Internet made it possible to publicize initial intentions and to promote new practices. This literature served to mobilize both designers and users.

NOTES

1. On the history of the telegraph and television, see Flichy (1995).
2. Witness account by William Preece, future technical director of the British Post Office. Quoted by Marvin, 1987: 206. On the beginnings of electric light in France, see Beltran and Carré (1991).
3. Lewis Mumford was also particularly interested in technological utopias. See Mumford (1967).
4. Louis Sébastien Mercier, the author of *Tableau de Paris*, is often considered to have written the first science fiction novel, *L'An 2440*, published in 1771, in which electricity played a key part.
5. Tarde's novel, although published at the end of his life, was written in 1879, before his sociological work. It is therefore a sort of sociological story which prefigures some of the themes he was later to address in his social science work.
6. See, in particular, Flammarion (1911) (chapter "Un grand esprit méconnu: Charles Cros', pp. 479–83).
7. In 1902, the Académie française offered the Guzman prize to the first person to establish contact with beings on another planet. Mars was excluded from the prize because it was believed that this planet was without doubt inhabited.

8. See, in particular, Jacques Sadoul (1984: 29–30).
9. The link between the railway and democracy is also one of the elements of Saint-Simonian thinking in France. Michel Chevalier (1836: 3) wrote, for example: 'Improving communications means working towards true, positive and practical freedom ... It means extending the freedom of the greatest numbers as far and as well as possible, and doing so through the law of the vote'.
10. Mumford was one of the main disciples of Geddes. In his history of technology, electricity is at the heart of the neo-technological phase. See Mumford (1934).
11. On the association between the coal and metallurgical industries and hell, see Williams (1990: 66), who cites numerous literary texts describing the industrial world as a hell. She also gives the opposite example of illustrations of the 'paradise lost' which refer to a mining landscape.
12. This point, which may appear hagiographical, is confirmed by several authors and particularly by Rogers and Larsen, 1984 and Young, 1988.
13. Many texts showed, for example, the computer user leaving his or her body to live a new disembodied life in different virtual worlds. Others presented cyberspace as a social space completely independent from 'real' society, capable of regulating itself without any intervention from either the state or even the market. Finally, others imagined that the circulation of data in worldwide networks would completely change the economic rules. Firms of the old economy would disappear like dinosaurs and the many shareholders of dot.coms would become millionaires.
14. With regard to hackers' ethic, Levy notes, for instance, that 'people who trotted in with seemingly impressive credentials were not taken seriously until they proved themselves at the console of a computer' (Levy, 1985: 43).

7. The birth of socio-technical frames

Following the presentation of the work of historians of innovation, including economic historians of technology and historians of the *imaginaire*, we are able to draw some conclusions. The history of technology and its uses is constructed along three entangled lines: that of chance, that of human necessity and that of human will, both individual and collective. As indicated above, historical causality is profoundly different from that of the natural sciences, being more a matter of judgement than of deduction. As Paul David has shown, among the multiple causes which might explain a technological event we find a whole series of possible choices which have become irreversible. The higher upstream we are in the innovative process, the more open the choices will be. The solutions opted for depend on chance and the action of the strategic players. The lower downstream we are, the more closed the choices will be, for they are determined by the past. A complex technological device will involve some elements for which the choice is open and others for which it is more closed.

In a situation of open choices, as in the artistic activity studied by Michael Baxandall, one must not consider that strategic actors are influenced by a particular (technological or social) phenomenon, but rather that they will mobilize a number of resources in making their decisions. It is this historical perspective of open technological choices that I use to study the birth of socio-technical frames. When a frame of reference is developed, other alternatives exist. Explaining these choices means being able to analyse the alternatives which were not chosen and explaining why that was so.

In this chapter I start with the distinction between frame of functioning and frame of use formulated in Chapter 4, and successively study their birth and articulation within the socio-technical frame. Finally, in a concluding section, I revert to the major steps characterizing the development of an innovation.

FROM TECHNOLOGICAL *IMAGINAIRE* TO FRAME OF FUNCTIONING

Consider, first, how a frame of functioning is created. As indicated in the preceding chapter, there is no continuity between a technological utopia and the realization of the first artifacts. The utopia may have its own development without triggering off even the slightest beginnings in the design of an artifact. To create a technological product, the inventor has to place his or her project in a frame of functioning. For example, to fly, the inventor might think of developing a machine lighter than air. However, to travel with that machine it will have to have a locomotive device (engine and propeller). Here then are two possible frames of functioning which lead to an airship. But the use of an engine also makes it possible to fly a machine which is heavier than air. Two opposing options are thus present here. We note that this elementary, even simplistic summary of the development of flying machines falls squarely under the criticism levelled in the preceding chapter at the *a posteriori* view of technological history. Joseph de Montgolfier did not reason in terms of 'lighter than air' since he attributed climbing power to smoke, and at the end of his life imagined that it was electrical phenomena which caused the balloon to ascend (Gaudin, 1984: 21). A frame of functioning is therefore not the encounter between a technological utopia and a scientific theory; its formulation is far more complex.

At this stage of the innovation process inventors may successively try several frames of functioning. For example, from the start of the electric telegraph the community of telegraphists and electricians imagined wireless communication. Then, throughout the nineteenth century several frames were tried, for instance, using the earth or water as a conductor, or electrostatic and then electrodynamic induction. But the technological projects which try to create a utopia by systematically using available scientific knowledge are not always successful. It may also happen that an innovator imports a scientific theory which a priori does not solve the problem yet which, contrary to all expectations, is successful. Thus, whereas physicists considered that microwaves had a maximum range of less than one hundred metres, Marconi believed that they were the artifact that would make wireless telegraphy possible (Aitken, 1976).

A new frame of functioning may also be formed through the transformation of an existing one. Alexander Graham Bell, for example, wanted to use the frame of functioning of telegraphy as a means to achieve a specific goal: the creation of a multiplex device which would transport several messages simultaneously. During his experiments he noticed that a telegraph wire could transmit sound, and consequently constructed a new frame of functioning, that of telephony.

These different operations in the formulation of a frame of functioning very often involve diverse procedures of transfer, movement, mediation and translation which may relate to science or another frame of technological functioning. In his research on the birth of the Citroën 2CV, Patrick Fridenson (1988) showed that part of the originality of this car derived from the transfer, to the vehicle, of materials and technical solutions used in aeronautics. Those in charge of the design department of the firm had personal experience in aviation where they had found a solution for decreasing the weight of the new vehicle – the condition *sine qua non* for designing an inexpensive car for the working classes.

Transfers may also be made with science. The shift from one field to another may be from science to technology or technology to science. In the former case it is generally considered that the engineer applies science, and in the latter that the scientist tries to explain the functioning of technological devices. Technological devices may also be applied as scientific instruments, especially in technoscience today, and transform science.

Let us revert to an emblematic case in the nineteenth century. In his conclusion to a remarkable study on the birth of the radio, US historian Hugh Aitken considers these operations of transfer which are more than a mere exchange of information. There has to be an active agent who takes care of the transfer, the mediation. Some inventors and particularly those at the origin of new frames of functioning fulfil this social function: 'They are translators. They must be fluent in more than one "language", at home in more than one world, adept at playing by more than one set of rules' (Aitken, 1976). These translators thus end up creating a new language. This problematic of translation, defined 30 years ago by Aitken and more recently adopted by the French sociology of translation (Latour, 1987) is obviously essential to an understanding of how a frame of reference is created.

But the creation of a frame of functioning is not limited to the choice of a few grand principles. Gilbert Simondon's analysis of the process of materialization (see Chapter 5 above) clearly describes the thankless daily task of the designer. A technological device becomes stable only when a definitive form has been found for the arrangement of the various parts. In the initial stages of an innovation there are principles, plans and abstract descriptions of processes. Later there are mock-ups and prototypes which enable the designers to verify the original intentions. Of course the artifact will be different from its first version on paper; the encounter with the resistance of the materials and objects will change the initial ideas. The first object is merely 'the physical translation of an intellectual system' according to Simondon (1989: 46). Later, the different elements will converge in a concrete, more compact technological object. Each element will then be integrated into a system which has acquired its own coherence. The object will have become a 'black box' which works, both for the uninitiated and for the engineer, without the user having to be familiar with the articulation of the different component parts. The frame of functioning will thus have found its stability and the phenomenon of technological lock-in will occur.

A technological object is never isolated. It is part of either a family of innovations with the same technological components, or a broader technological system. For example, data-processing techniques constitute the basis of digital telecommunications and are also used in video images. Television is another technological system. The advent of colour TV in the 1960s and the development of digital TV today both involve the necessary, simultaneous invention of cameras, systems of transmission and receivers. This systemic aspect is an important component in the frame of functioning which constitutes the shared mode of functioning of the different technological tools.

The frame of functioning also organizes the interactions between the different actors. As in the paradigms of Kuhn, Constant and Dosi, it ends up being adopted by an entire technological community. This choice then stretches out from the world of R&D to that of production or maintenance. It also frames the users' actions. Of course this frame of functioning does not spread directly from laboratories to users without any change; it is negotiated with the different actors, in various ways. It may result either in a frame of functioning which is strictly identical for all the actors, or in compromises which enable each actor to find what he or she wants. The negotiation of international standards for network technology (electric energy, telecommunications) is a good example of the development of a common frame of functioning. In order to interconnect networks and terminals, it is necessary to adopt common technical standards. Due to the increasing complexity of networks and competition between operators, agreement is not always reached. In such cases a compromise is established on a group of standards with a common trunk and multiple variants which may be common to a subset of actors (see David and Foray, 1994; Hawkins, 1993; Bucciarelli, 1994: 165–74; Marti, 1994). Coherence and diversity are thus articulated.

Negotiations on the frame of functioning take place not only in the context of standardization institutions, but also at various levels starting with that of the laboratory, between its members, and then between the laboratory members and their technological community in the form of articles or papers delivered at conferences. Laboratories also have to negotiate with producers. The methods department of a factory might, for example, refuse a proposal or ask for an adjustment in order to be able to produce the object efficiently and cost-effectively. A final negotiation, with the users, concerns the technological representation that will be displayed and the interface between user and machine.

There are numerous types of negotiation with users over the frame of functioning. At the laboratory stage, users are already taken into account. For example, in the design of an audiovisual terminal the space it occupies and the visual and sound environment of future users are considered to be relevant factors. The marketing division also intervenes to adapt to the technological representations of future users. The ergonomics department studies the interfaces. Each section, through different tests, negotiates with real individuals, who are meant to represent the mass of users.

In short, the development of a frame of functioning concerns not only the innovators but all the actors involved in a technology. After negotiations in various arenas (see Chapter 4, above), the end result is a boundary object.

FROM SOCIAL *IMAGINAIRE* TO FRAME OF USE

The name of technological objects may relate to their functions or to their use. For example, the same data-processing machine was called both a microcomputer and a personal computer. The inscription of its uses in the name of a technological artifact is a sign that a certain type of use is associated with a certain type of object. Alain Gras points out: 'in order to understand the success or failure of a technology, it is not enough to write its internal history. The technological deed ... has something to say to the imagination; it is inscribed in the collective consciousness and finds its meaning there while simultaneously participating in the transformation of the environment' (Gras, 1993: 83).

This inscription of the technological fact in a broader social history appears clearly when one examines the imagined uses of a new technology. For example, between the mid-eighteenth and mid-nineteenth centuries, three types of use of telecommunications appeared. Texts and engravings from the eighteenth century generally associated the semaphore telegraph or string telephone with romantic communication. At a time when intimate correspondence was fairly widespread among the aristocracy, it was this frame of use that was imagined for the new communication tools. During the French Revolution and subsequently under the Empire and the Restoration a new, more structured and unified state emerged and telecommunication was imagined to be a new means of state control. For Bertrand Barère, the telegraph was a means 'to consolidate the unity of the Republic by the intimate and immediate contact which it gives to all parties' (quoted by Flichy, 1995: 9).

Finally, in the 1830s a new frame of use was envisaged for the telegraph: that of the transmission of stock market rates. This use was not only proposed by the promoters of new telegraph lines, it was also recognized by the opponents of these projects who came either from the political class or the business world. Thus, in 1837 a member of the French parliament,

Jean-Claude Fulchiron, considered that telegraph lines for private use served only to 'establish brigandage, so as to rob those who do not have news of the Paris Bourse' (Flichy, 1995: 23). During the same period, the Belgian chambers of commerce considered that 'the telegraph is useless for commercial purposes trade and serves the sole interests of stockbrokers' (Vercruysse and Verhoest, 1994: 97). It is important to note that this evolution of frames of use occurred independently from another debate that was to appear a few years later between the partisans of two frames of functioning: the semaphore and the electric telegraph.

Even if the question of the frame of use relates to the main representations of an era, it is nevertheless significant that this frame, like the frame of functioning, is the outcome of a complex process. Initially there is a confrontation between the different representations of the new technological artifact. Then, as in Simondon's schema, a phase of materialization occurs. Finally, the first users alter the frame of use and a phenomenon of feedback is witnessed in relation to the initial frame.

Let us revert to the beginning of the process. Among the different representations of the technology, that of the engineers is interesting to observe. As François Caron notes: 'the technical literature of all ages, from engineering journals to patents for inventions, expresses a coherent and precise societal project which justifies the described invention, always presented as an answer to a need or aspiration' (Caron, 1991: 117). In these major societal projects the designers often chose a particular application enabling them to construct a first frame of use, essentially for demonstration purposes and to convince investors, their laboratory director or the corporate management, of the advantages of the new technological artifact that they were developing. Thus, Bell, wanting to demonstrate in public places what the telephone was, asked one of his assistants a few miles away to play a musical instrument. He was thereby proposing a frame of use for the telephone based on the transmission of music rather than an exchange of voices. Various historical documents suggest, however, that Bell saw this not as the frame of use of the future telephone but simply as a demonstrative frame. Similarly, in the mid-1990s the different laboratories working on the new broadband high-speed telecommunications networks with asynchronous techniques, regularly presented multimedia experiments in which images played a large part. In this way they wished to show the iconical possibilities of these new networks, while many of their promoters believed that they would initially serve to link up very high-speed computers.

When designers subsequently want to establish an effective first frame of use, they often tend to adapt that of similar objects. Thus, the first engineers of the telephone imagined a frame of use similar to that of the telegraph: the transmission of messages. Edison, a professional inventor working for the telegraph and telephone companies, also thought that the phonograph could be used to record telephone messages. A conflict then broke out between Edison and his licensees. The latter proposed another frame of use: the playing of music in public places. Other innovators such as Emile Berliner also had the idea of 'capturing' music and of transferring the phonograph to a radically new frame of use: that of the Victorian family which, being an avid consumer of music at home, might wish to adopt a music machine.

The first American engineers of the telephone attempted to formalize their view of its frame of use by publishing instructions in the front of telephone directories. These specified that calls were to be short so as not to block priority business calls. However, users, especially in rural areas, were to develop a different frame of use: that of telephone chatting (see Fischer, 1992b). We thus see the appearance of a new actor intervening in the formulation of a frame of use. It is important to be aware that this actor was not alone, for this frame of use was also the business of the marketing people and the inventors. Each of these actors imagined an abstract frame of use, to use Simondon's expression, and the negotiation spawned a concrete frame. This type of negotiation may have one of many forms. It may be direct. Claude Chappe, for example, discussed with the French parliament the frame of use of his telegraph (Flichy, 1995: 8). It may also be indirect. In that case it takes place with the virtual representatives of the users which the various opinion poll techniques help to reveal. Through the mechanics of the process itself, questioning users may be instrumental in constructing a frame of use. In the conclusion to an article on the history of television opinion polls in the 1950s, Cécile Méadel notes: 'the television of those years was built on the questions it put to its viewers. The mirrors which it constantly set up around itself drew its own profile, thereby enabling it to define itself by defining others' (Méadel, 1990: 53).

In this game of successive formulation of a frame of use, two significant places of mediation exist: the project team and the first sites where the artifact is used. The project team is relatively formalized in contemporary innovation programmes and is usually composed of the representatives of the different services of the firm (R&D, production, marketing and so on).[1] In the nineteenth century it was generally limited to one family (the Chappe brothers, the Lumière brothers) or to an inventor with a few assistants (Bell, Edison, Marconi). In this team the question of a frame of use is debated extensively during the period in which the main functions have to be defined, and later when the product is to be launched on the market. Within the group, each of the participants arrives with his or her own representation of potential uses and knowledge of lifestyles. From this comparison a first frame of use is usually born.

The first uses of a new technological system provide another opportunity for the frame of use to be transformed. Consider a contemporary example. Numerous sociological studies carried out on the beginnings of French videotext clearly show how the shift occurred from a conception of the videotext as an access to the major databases, to a new vision in which dialogue became essential either through message services or, more simply, in the context of transaction services.[2] Some authors have seen this development as the users' revenge on the engineers. It is, more simply, a normal phenomenon of mediation between designers and users.

After these different mediations which transform the initial frame, the new frame of use acquires a degree of stability. This frame can be compared to a social norm or convention. It may even have a legal form. The few cases that I have studied in the telecommunications and broadcasting fields clearly show the potential role of the law. It is used primarily to support a declining frame of use which, in a sense, refuses to make way for a new emerging frame.

From the time of the French Revolution and until the 1830s, the frame of use of the telegraph was clearly that of state communication. In 1832, private telegraphy was established for the purpose of transmitting stock market information, and a line was built between Paris and Rouen. Despite some reservations, the state decided to prohibit the project and parliament passed a law in 1837 instituting a public telecommunications monopoly. Fifteen years later a new law authorized the commercial use of public networks. The same thing occurred a century later, regarding French broadcasting. Although in the post-Second World War years the state developed radio and television in a framework of public monopoly, it was only in 1972 that the monopoly was legally codified, and in 1978 that violation of the monopoly was punished. Just when a new frame was appearing – a plurality of stations and programmes – the law codified the former frame of use (see Giraud, 1993).

THE SOCIO-TECHNICAL FRAME OF REFERENCE

An innovation becomes stable only if the technological actors have managed to create an *alloy* between the frame of functioning and the frame of use. As in any alloy, it is impossible to distinguish the initial components from the final product. The socio-technical frame is not the sum of the frame of functioning and the frame of use, but a whole new entity.

The articulation between the two frames does not correspond to a necessity; it is part of a range of possibilities. For example, technical devices for the transmission of sound may be used to transmit music or conversation. However, to construct new alloys, as in the case of radio which combined the transmission of sound by electromagnetic waves (maintained by means of a triode valve) and families listening to popular songs in the cosy comfort of their living rooms in the inter-war years, it took the particular history of individuals who bridged the gap between the different worlds. Among these intermediaries was Marconi who was to free radio waves from the amphitheatres where they served solely to illustrate James Maxwell's theories, and thus to make them a true means of distance communication. Reginald Fessenden, another such intermediary, simultaneously tried to create another device for the transmission and reception of radio waves, for the purpose of transmitting sound. Lee De Forest also endeavoured to find another way of transmitting sound, via a triode, and thus to broadcast information and music. Each of these individuals made a choice which helped to define and reorientate the new socio-technical frame. Gradually this merging, which at first appeared surprising and even unseemly to their contemporaries, developed a solid base. Manufacturers and users produced and consumed within this new frame of reference. Socio-technical lock-in finally occurred and a new alloy was created. Even if the articulation of radio waves and family music seemed random at first, at each stage of the mediation the randomness was reduced.

It is nevertheless essential to be fully aware that we are not witnessing a univocal model of technological and social determinism. Radio waves did not create family music, no more than family music invented radio waves.

Although there are many designers capable of articulating frame of functioning and frame of use, there are also 'innovation-users' who, with their community, perform this mediation. Eric von Hippel has shown a keen interest in these new intermediaries. He defines an innovation user as someone who has sufficient incentive to innovate, who willingly agrees to reveal his or her innovations and who can produce and distribute them at a low cost, so that they can compete with commercial networks (von Hippel, 2002). Open source software is a good example of innovation users who are in a strategic position.

There is a case that differs slightly from the one studied by von Hippel, and that is the Internet. The designers of Arpanet or Usenet were academics who used resources from ARPA or their university to create network computing. But they were also the first to use this network, with applications such as email, newsgroups and, later, the Web. Owing to government funding, the designers were able to use their network. This made it easier for them to link up frame of functioning to frame of use, for the cooperation underlying the design of the network was the very substance of that use (Abbate, 1999).

These different cases of intermingling frame of functioning and frame of use relate to the question of mediation. This approach to mediation (see

Hennion, 1993, 1995) was developed to a large extent by the contemporary English-culture school of the social history of art. In *Painting and Experience in Fifteenth Century Italy* (1972), Baxandall shows that painters' production is a compromise between their project of personal expression and the way in which they anticipate the demands of their sponsors and the expectations of their public. Both painter and public embed their relationship in a shared cultural repertoire which constitutes the mediation linking them. During the Italian Renaissance this was above all the Bible, taught by preachers: their sermons 'coached the public in the painter's repertory, and the painters responded within the current emotional categorization of the event' (Baxandall, 1972: 55). Thus, 'preacher and painter were *repetiteur* to each other' (ibid.: 49). Mathematics used by merchants to measure quickly in a world with no unified system of measurement constituted a second type of mediation. The painter 'made pointed use of the repertory of stock objects in the gauging exercises': cisterns, columns, paved floors and so on (ibid.: 87). Finally, 'the harmonic series of intervals used ... by painters was accessible to the skills offered by the commercial education' (ibid.: 101).

Baxandall described his method in his book *Patterns of Intention* (1985). He was careful to steer clear of hasty comparisons between art and other social phenomena. For example, he criticized the analogy proposed by some between the first Cubist painting of Pablo Picasso and the new physics of Albert Einstein. This type of comparison overlooks an analysis of the mediations between art and science. According to Baxandall, in order to talk of analogy two conditions must be met. First, science must pay attention to the specific aspects of the visual experience of which a pictorial equivalent may perhaps be found. Second, it is necessary to find some signs proving that this type of comparison was at least conceivable for the people of the day (Baxandall, 1985: 76).

In a previous study (Flichy, 1995: 17–31), I performed the same type of analysis concerning the Chappe telegraph during the French Revolution. I tried to reconstruct the mediations which were set up to ensure that the semaphore and state communication formed an alloy. The main locus of mediation was the Public Education Committee of the Convention to which various proposals were submitted by scientists and engineers. This committee was also responsible for the reform of the calendar and weights and measures, and some of its members had already participated in the administrative partitioning of France. It was therefore the driving force behind a vast undertaking of rationalization which primarily involved the partitioning of time and space. The committee assessed the Chappe project and realized that it had the means to conquer time and space and to guarantee the unification of the Republic. Hence, it was the rationalization of time and space and, more generally, Enlightenment thought which provided the mediation between the semaphore and state communication.

In the creation of the alloy between frame of functioning and frame of use, an important mediator intervenes: the price. Clearly, we are not in a context of microeconomic theory where a demand curve (quantity depending on price) meets a supply curve. In the situation of interest to us here – the creation of frames of use and functioning – there are usually too many uncertainties for such curves to be drawn. It is nevertheless possible to know whether a degree of correlation can exist between the price of the frame of functioning and that of the frame of use. The pioneers of microcomputing referred to earlier on were unable to create the home computer which they had in mind because the cost price of microcomputers was too high to allow for domestic use. It was necessary to change the frame of use and to aim for a personal computer for office use. A part of the initial utopia is nevertheless found in this tool which has often become an instrument of autonomy compared to computer mainframes. Finally, price mediation eventually transformed the initial socio-technical frame. Today the decline in cost prices has made it possible to revert to the initial project for a home computer.

At the intersection between the two frames we also find the external aspect of the technological object and the points of contact between humans and machines, what in computer language is called human–machine interfaces. The external aspect of a new object often has a lot to say about the tension between the frame of functioning and the frame of use. Take, for example, the first railways. Konrad Lorenz notes:

> [A]t first it seemed enough to put a stage coach on rails. Then it was found that the wheelbase of the horse carriage was too short, so it was extended and with it the entire coach. ... As strange as it may sound, a whole series of usual bodies of ordinary stage coaches were placed one after the other. These coaches merged at the sides and became compartments, but the side doors ... remained unchanged. (Lorenz, 1976: 40)

This process in which the volume of the wagons was established, clearly indicates a continuity in the frame of use with the stage coach. By contrast, in the frame of functioning there was a total break. As Schumpeter notes, no matter how much one changes the stage coach, you will never end up with a railway.

If we consider a contemporary object such as the microcomputer, the interfaces will obviously differ, depending on whether the machine is intended essentially for office or home use or specifically for games (gaming platform). Thus, for the same basic object, a different frame of use will change the object only marginally.

We note that during the history of the development of a socio-technical frame, the role of the frame of functioning and the frame of use is never the same. In some cases, generally new products, the materialization of the frame of functioning precedes that of the frame of use. In the laboratory, representations of uses are often vague. More thought is given to abstract than to social use. By contrast, in the case of new processes the frame of use is defined first. Thus, in the case of the scientific equipment studied by von Hippel (1988), users define the frame of use precisely and designers work on that definition.

Finally, there are cases in which the articulation between frame of functioning and frame of use is complex and requires the intervention of a particular mediator. Major corporate computerization projects are a fine example. Computer departments often mediate between manufacturers and users. They choose the different hardware and software required and the functions proposed to users. In the most complex cases, such as enterprise resource planning (ERP), they entrust consultants with the task of parametrizing software (Segrestin, 2004). They also train and help users. In short, they transform the abstract use of manufacturers into a prescribed use. Users then try to adjust this prescribed use to shape a real social use.

Once the socio-technical frame has been stabilized the actors will consider the frame essentially as a black box. They may, however, intervene actively on a specific element: laboratories and manufacturers on the frame of functioning; marketing services and users on the frame of use. This creates a situation in which the interaction between the two frames is weakened.

To conclude this reflection on the creation of a socio-technical frame, we note that the process of stabilization is slow. At first the frame is still extremely fragile. The formula for the alloy has not yet been perfected and is easily altered. Gradually the community of technicians agrees to the new frame. The supply is diversified as other manufacturers appear. The marketing and maintenance people are convinced of the advantages of the new frame, with the result that more and more users adopt it. As the alloy becomes more solid, alternative frames are abandoned. This is the beginning of a 'virtuous dynamic' of increasing returns to adoption (see Chapter 5, above). There is a combination of the effects of learning through productive practice and economies of scale, and of the effects of learning by using, external network economies, increasing returns on information and technical complementarities. Eventually the socio-technical frame is set and we have what is known as 'technological lock-in'.

The stable socio-technical frame concerns not only a technological artifact but also an entire technological system. Thomas Hughes's work on the early days of electric lighting clearly shows, for example, that it was necessary not only to combine a source of energy with an electric bulb, but also a system of transport with principles of pricing (Hughes, 1979).

THREE PHASES OF INNOVATION

In the process of creation of socio-technical frames, several phases found in most innovation can be distinguished. First, in a moment that could be characterized as the prehistory of innovation, various parallel and unconnected histories take place. In a second phase, several elements start to converge in a way which is still utopian and abstract; this is the catchall-object phase. Finally, in the third phase, different actors are brought into contact; they negotiate and reach an agreement. This is the boundary-object phase.

Concurrent History Phase

In the study of the prehistory of an innovation, that is, the period preceding the establishment of a socio-technical frame, a question of method arises: what are the social worlds that the historian of innovation has to summon? A priori, two alternatives exist. Either historians limit themselves to the choice of actors, or they select the social worlds which seem to them to be important, given the final form of the innovation. In the first instance there is a risk of excluding a social world which may play an essential part later in the development of the innovation but to which the first innovators paid little attention. For example, when Edison developed the phonograph he showed no interest in using it to play music. Therefore, there seems no reason to study the evolution of the playing of music at home.

If we opt for the second alternative, there is the chance of us producing a 'retrospective illusion of inevitability'. The historian is then in the position of a novelist who writes the history of different characters destined to meet one another. Moreover, he or she will exclude the particular social world of an innovator who was subsequently eliminated. In the example of the phonograph, the historian will pay little attention to office work, although for Edison that was the frame of use of 'the talking machine'.

The only way of solving this contradiction is by writing the concurrent history of all the social worlds concerned: those summoned by the innovators as well as those which appeared later on and were subsequently to play an essential role. In the case of the phonograph several worlds were involved in the establishment of its frame of functioning, like those of the specialists of telephony (Edison, Bell, Berliner and so on) or the scientists interested in recording speech (Scott de Martinville). With the frame of use the diversity was even wider. Several parallel histories need to be studied: that of office work which was starting to be automated, that of telephone operators who wanted a relay memory for long-distance communication, that of all the collectors and other souvenir-lovers and, of course, that of music-playing of all kinds which developed around the piano and singing (Flichy, 1995).

These parallel histories are of different orders. Those related to the frame of functioning belong to technical communities: small, specific groups of people. By contrast, those histories associated with the frame of use concern far broader social groups; they are situated more in the tradition of the French historiographic *École des Annales* and the history of mentalities and habits – a history of the long term. Finally, it is important to note that with the variety of these parallel histories, an innovation does not have a single point of origin; on the contrary, it is rooted, to varying degrees, in diverse fields.

The Catchall-Object Phase

The encounter between the different social worlds initially occurs in an imaginary mode; that is the catchall-object phase. On the one hand, writers or journalists imagine new technologies and their applications; on the other, inventors propose numerous uses for their machine, to convince their sponsors and society at large of their social utility. The potential actors of a new technological object provide projects and utopias which may concern both a mode of technical functioning and a new use.

This abundant utopian literature is found to a large extent in the field of network technology (transport, energy and so on) and even more so in that of electronic communication. The contemporary example of information highways has been this catchall-object phase. Consider this example in more detail. Several actors from the industrial and political worlds have had partly convergent utopias that have nevertheless led to different projects. For industrial companies, information highways are perceived as a challenge to the boundaries between telecommunications, the broadcasting media and computers, that is, as technological convergence. The computer industry sees its project as an opportunity for launching a new range of computers. Software manufacturers prepare related programs and also offer their know-how for interactive television or to provide users with intelligent management of the multiple programmes offered on digital TV. For network operators, information highways are also becoming an opportunity to diversify. Some cable operators propose a telephone service routed via the cable network, while telecommunications operators are testing the possibility of broadcasting TV programmes via their networks, to offer online data services.

If we switch from technology to politics, we note that some governments see information highways as new infrastructures which could serve as a basis for economic revival if they acted as an essential component in a Keynesian policy of major public works. For other politicians these highways are, on the contrary, a way of putting an end to rigid regulatory frameworks. A dynamic liberalization policy should lead to price reductions in the telecommunications and broadcasting sectors, and hence to strong growth in consumption.

Finally, some potential users of telecommunications services see information highways as affording an opportunity to revive projects for specific uses which have never reached maturity. National development agencies are reconsidering their projects for telework. Likewise, tele-health and tele-education seem to some to be a good solution for checking or even stabilizing expenses in the health and education sectors, which are growing far faster than the GDP.

Thus, the hopes placed in information highways largely exceed what a technological object can offer, no matter how open it may be. Above all, the projects envisaged are at best juxtaposed and sometimes even clearly opposed. The catchall-object period is therefore particularly unstable; either it exhausts itself or it evolves into a phase of negotiation and development (Flichy, 2007: 17–34).

The catchall-object phase also reflects the profoundly ambiguous nature of technology highlighted by Cornelius Castoriadis:

> The awe inspired by artifacts, the facility with which the man in the street and the Nobel prizewinner alike allow themselves to be imprisoned in new mythologies ('machines which think' or 'thinking like a machine') often go hand in hand with rising protests, by these same people, against technology, suddenly blamed for all humanities' problems. (Castoriadis, 1992: 123)

If we interpret this entire catchall-object period as an ideological balloon which will subsequently be deflated when the promoters of the new technology are faced with the hard realities of production and marketing, it can be seen as the period in which different converging elements, which might later constitute socio-technical frames, are being outlined in an imaginary mode. It is in this period that engineers discover unimagined possibilities for use, and that users hear about a new technology which they never imagined existed.[3] Thus, at the beginning of the 1990s information highways became a topic of public debate launched primarily by Al Gore, and in 1993 the Internet became the new avatar of information highways (Flichy, 2007: 31–4). But the intense public debate that preceded these events enabled the Internet to leave the closed world of computer specialists, and opened new vistas in the media and telephony fields.

The Boundary-Object Phase

The boundary-object phase corresponds to one of indeterminacy in technological options. A wide range of possibilities remains open, with respect to both the frame of functioning and the frame of use. It is therefore necessary to remove the ambiguities, to clear the confusion and to define an object with a bolder outline; to move on from utopia to reality, from abstraction to materialization, and to construct a boundary object. In this transformation from a catchall to a boundary object, the different projects have to be sorted and only the potentially compatible ones selected. But this compatibility is not a foregone conclusion; it has to be constructed, negotiated with the different actors concerned. These actors will be able to participate in a common project only if they consider it to be in their interests. On the other hand, not everything is open to negotiation; each world has a certain number of specifications which are at the core of their functioning and which the other partners cannot challenge. In this sense, my notion of a boundary object is close to that of the interactionists (Star and Griesemer, 1989). The process nevertheless differs slightly in so far as there is a shift from vague projects with a strong utopian component to realistic, concrete and operational projects. In other words, promises are fulfilled.

Before any negotiations it is the controversies and confrontation which appear; not only at the level of discourse, as in the catchall-object phase, but also at the level of technical creation. Alternative current is opposed to continuous current, the gramophone to the phonograph, digital calculators to analogue calculators. The conflict affects not only the frame of functioning but also the frame of use: the use of calculators for management is opposed to military use; the recording and listening to sound can be used at work or at home and so on. Each of these conflicts corresponds to a partial understanding of the technological object, to a specific element of its functioning or its use. Therefore their resolution cannot be local; it has to be organized globally. In the early twentieth century, debates in the US on the radio-telegraph and the radio-telephone, on the one hand, and on electronic valves (diodes, triodes) on the other, were independent yet were resolved together. The triode was a key element in the development of radio, but at the same time the expansion of radio favoured the development of this valve. Similarly, the gramophone (disk) triumphed over the phonograph (cylinder) because its promoters banked on a musical use and not on an office machine.

In this process of creation of a boundary object based on different social worlds, certain individuals play an important role of mediation. Berliner linked the recording of sound to family music, Edison linked electricity to lighting, and so on. Such mediation requires adequate knowledge of the different social worlds concerned, in order to find a solution acceptable to each one. It may correspond either to a compromise (as in the case of the launching of the compact disk by Philips and Sony) or to a 'capture' (Marconi used radio waves, a laboratory object, to make the wireless telegraph). At the end of this mediation the socio-technical frame is set and there is socio-technical lock-in. It is often in the organization of financial channels between the different actors involved in innovation that the conventions of cooperation appear most clearly. In the case of the Minitel, France Télécom is paid by users and then distributes the revenue between the two networks concerned (telephone and Transpac data network) and the server (computer) centres which, in turn, pay a part of the income to the information providers (particularly the press). This circuit shows that the videotex is no longer the child of the telecom operator but a boundary object on which different partners work, including the press which was initially opposed to it (Jouët et al., 1991). In the 2000s, Google found another compromise. It launched its search engine, without departing from the Internet tradition of free information, on the basis of an efficient business model. Google provides a high-quality service, based on a mode of classification in terms of number of citations, derived from academic bibliometric rules, and on a simple and effective portal. In addition, and graphically separately, with AdWords it proposes an effective advertising system that is self-regulating. Advertisers choose the words on which their advertising links will appear, as well as the rate that will set their order of classification in the advertising part of the page (Battelle, 2005).

STABILITY OF THE SOCIO-TECHNICAL FRAME

The new boundary object is very different from the initial catchall object. It would therefore be meaningless to consider the catchall object as a matrix of the final object. The latter is the result of constraints derived from a far broader frame of reference (telecommunications for the videotext), from chance convergence, and from the action of strategic players. Innovation will continue to develop within the new socio-technical frame. Different ranges of technological objects will come into being and several generations of materials or services will emerge. Uncertainty is nevertheless weaker; we are in the realm of tactical innovation.

For the producers, this stability squares with the management idea of 'dominant design' (Abernathy and Utterback, 1978). The innovative activity is directed to improve the process and not to explore new alternatives. For users, they 'poach' in the socio-technical frame, appropriating and reappropriating the new object. In this respect more classical forms of analysis are found. As Pierre Dockès and Bernard Rosier note, 'when the repetition is sufficiently frequent for the "common knowledge" to be stabilized (everyone knows that the others know that he knows), the singularity disappears and the story makes way to traditional economic theory' (Dockès and Rosier, 1991: 201). Economic calculation then becomes possible and it is easier to evaluate demand. Production costs are calculated with a fair amount of certainty. In this innovative system, cost reduction is an integral part of the designers' challenge. Their main preoccupation is often to find technical solutions to create a product at a specific cost.[4]

In this period of innovation, forecasting is possible. By contrast, preceding periods are ones of retrospection. No forecast can be made with even the slightest degree of certainty; the diversity of technological utopia is there to prove it. The coherence of the first three phases is constructed by the strategic actors on the one hand, and by the historian on the other, and these constructions are fundamentally different. The former is situated in terms of projects and opportunities grasped; the latter is reconstructed from a final result which, by definition, is unknown to the actors.

NOTES

1. See, in particular, Midler (1993); Bucciarelli (1994). See also Guterl (1984).
2. See, in particular, Ancelin and Marchand (1984); Jouet et al. (1991).
3. A similar analysis has been carried out in the scientific field, on the case of 'membrane technology' (Van Lente and Rip, 1998).
4. Christophe Midler (1993: 26), who studied a new car project at Renault, noted the following statement by the project manager: 'The fundamental question is: what can we do for this budget?'.

Conclusion

The difficulty in studying innovation stems from the fact that it is always based on tension between continuity and discontinuity. I have used the concepts of frame of reference and boundary object to illustrate this fact. Yet, and I think I have made this clear throughout the book, the analysis of innovation is itself a boundary object. The approach I propose is therefore situated at the intersection between an interactionist sociology of technology, a socio-technical history and an economics of technical change. If, as Marcel Mauss wrote, 'the unknown lies at the boundary of disciplines' (1985: 365), we need to create a boundary social science to account for such a complex phenomenon involving widely diverse actors.

Yet even if this science did exist, it would still be difficult to present all the dimensions of innovative devices. The approach proposed in this book is built around the relationship between technology and its uses, and around the study of individuals' socio-technical action. It is a little less concerned with the question of technological co-operation in the context of large organizations, whether these are R&D laboratories or industrial production plants. This is because the book does not claim to propose a universal approach to innovation; it is a boundary object among others, which has attempted to compare and articulate several social science disciplines and some fields of observation around the question of technology and innovation.

Can these elements of an innovation theory be of use to engineers and designers in the development of new technologies? More broadly, can these reflections help our society to master technology? No simple answers exist to these questions, often put to sociologists. In so far as it is actors who interact to construct a new artifact, within a set of given constraints, it is clear that neither sociologists nor anyone else can provide solutions for successful innovation. Solutions do not exist prior to innovation; they are constructed while the technological action is under way.

If we agree that technological objects are the result of three elements – the activity of actors, chance, and socio-technical constraints – social science research can be seen to have a dual function. First, it can help to identify the socio-technical frames in which the action is situated; second, the past cases of innovation studied all act as examples for future innovation. A multitude of social worlds are concerned by innovation. Very often designers are barely, or badly, informed as to their characteristics or may even be totally unaware of the existence of a particular social world that the technological object will inevitably encounter. I have shown, as an example, that 25 years ago the videotext, like the Internet today, was the fruit of a confrontation–negotiation among telecom operators, computer companies and media publishers. Clearly, these different social worlds are largely unaware of one another. Sociologists of innovation can promote their mutual awareness by facilitating mediation. While social science researchers can offer neither theories nor recipes for successful innovation, they can provide the relevant actors with resources for the creation of a boundary object. Just as geographers provided explorers with maps on which certain areas remained empty, yet which were a help in the progression into the unknown, so too can sociologists provide engineers with useful information for their journey through innovation.

Historians have a considerable advantage compared to actors. They know the outcome of particular socio-technical adventures. This work of explanation, consisting in linking together a series of singular events (by taking into account occurrences which had been forgotten), cannot be applied to the future. Unlike the natural sciences where there is symmetry between explanation and prediction, in history there is a break between the two.

Thus, from the numerous cases of innovation studied by historians it is impossible to derive a method for the future. The different cases will be used to motivate innovative action, but they will not be seen as constituting a grammar of situations of innovation from which the designer merely has to select a model corresponding to the given circumstances. Just as generals cannot win a war by fighting the battles of Iéna or Austerlitz again, so too it is not by wanting to reproduce the strategy of electric lighting or the Internet that innovators will be successful. On the other hand, they may be able to learn from past examples, which could allow them rapidly to perceive a social world in which negotiation is indispensable, and to grasp an unexpected opportunity more successfully.

Sociology has always striven to explain the mechanisms which ensure the stability of interaction between individuals and their relations to their world. How can the randomness of behaviour be reduced and different behaviours articulated? The sociology of technology has introduced the technological object into such thinking. We have seen the complex mechanisms which allow new artifacts to appear, and have examined how actions become predictable once the frame of reference has been established. Strategic action then gives way to tactical action. In the world of tactics, the predictability of action is enhanced and the constraints bearing on actors increase. As we have seen, these constraints are related to functioning or use.

To sum up, the innovative process consists in stabilizing relations between the different components of an artifact, on the one hand, and between the different actors of the technological activity, on the other. The sociotechnical frame governs the different relationships and makes it possible to adjust individual actions. Contrary to a long-standing idea, innovation is not the sum of a brilliant discovery and a process of diffusion. Rather, it consists in the coming together of parallel histories, successive adjustments, confrontations and negotiations, and the reduction of uncertainty. This process of stabilization concerns the functioning of the machine as much as its use, and its manufacturers as much as its sellers. The challenge in the sociology of technology is to establish how the social link is constructed in and by the machine.

Bibliography

Abbate, Janet (1999), *Inventing the Internet*, MIT Press, Cambridge, MA.
Abernathy William and Kim Clark (1985), 'Innovation: mapping the winds of creative destruction', *Research Policy*, **14** (1), 3–22.
Abernathy, William and James Utterback (1978), 'Patterns of innovation in technology', *Technology Review*, **80** (7), 1–9.
Abramson, Albert (1987), *The History of Television, 1880 to 1941*, McFarland, London.
Aitken, Hugh (1976), *Syntony and Spark: The Origins of Radio*, John Wiley & Sons, New York.
Aitken, Hugh (1985), *The Continuous Wave: Technology and American Radio 1900–1932*, Princeton University Press, Princeton, NJ.
Akrich, Madeleine (1989), 'La construction d'un système socio-technique. Esquisse pour une anthropologie des techniques', *Anthropologie et sociétés*, **13** (2), 31–54.
Akrich Madeleine (1992), 'The description of technical objects', in Bijker and Law (eds), pp. 205–24.
Akrich, Madeleine (1993), *Raisons pratiques*, **4**, 35–57.
Akrich, Madeleine (1995), 'User representations: practices, methods and sociology', in Rip et al. (eds), pp. 167–84.
Akrich, Madeleine, Michel Callon and Bruno Latour (1988), 'A quoi tient le succès des innovations', *Annales des mines*, no. 11, June: 4–17, trans. English (2002), 'The key to success in innovation', *International Journal of Innovation Management*, **6** (2), 187–225.
Akrich, Madeleine and Bruno Latour (1992), 'A summary of a convenient vocabulary for the semiotics of human and nonhuman assemblies', in Bijker and Law (eds), pp. 259–64.
Alpers, Svetlana (1983), *The Art of Describing*, University of Chicago Press, Chicago.
Alsène, Eric (1990), 'Les impacts de la technologie sur l'organisation', *Sociologie du travail*, no. 3, 321–37.
Alter, Norbert (1983), 'L'effet organisationnel de l'innovation technologique: le cas de la télématique', PhD in sociology, Institut d'Études Politiques de Paris.

Amblard, Henri, Philippe Bernoux, Gilles Herreros and Yves-Frédéric Livian (1996), *Les Nouvelles Approches historiques des organisations*, Le Seuil, Paris.
Ancelin, Claire and Marie Marchand (1984), *Télématique: promenade dans les usages*, La Documentation française, Paris.
Arnal, Nicole and Alain Busson (1993), 'Vers de nouvelles pratique audiovisuelles', *Réseaux*, **60**, 142–50.
Aron, Raymond (1957), *Introduction à la philosophie de l'histoire. Essai sur les limites de l'objectivité historique* (1st edn, 1938), Gallimard, Bibliothèque des idées, Paris, trans. English (1976), *Introduction to the Philosophy of History: An Essay on the Limits of Historical Objectivity*, Greenwood Press, Westport, CT.
Arrow Kenneth (1962), 'The economic implications of learning by doing', *Review of Economic Studies*, **29**, June, 155–73.
Arthur, Brian (1988), 'Competing technologies: an overview', in Dosi et al., pp. 590–607.
Baczko, Bronislaw (1978), *Lumières de l'utopie*, Payot, Paris, translated in English (1989), *Utopian Lights: The Evolution of the Idea of Social Progress*, Paragon House, New York.
Ballé, Catherine and Jean-Louis Peaucelle (1972), *Le Pouvoir informatique dans l'entreprise*, Les Editions d'organisation, Paris.
Barbichon, Georges and Serge Moscovici (1962), 'Modernisation des mines, conversion des mineurs', *Revue française du travail*, July–September, 1–201.
Barras, Richard (1986), 'Towards a theory of innovation in services', *Research Policy*, **15**, 161–73.
Baszanger Isabelle (1992), 'La tradition interactionniste et la sociologie des sciences et des techniques', in Pestre (ed.), pp. 35–69.
Battelle, John (2005), *The Search: How Google and its Rivals Rewrote the Rules of Business and Transformed Our Culture*, Portfolio, New York.
Baxandall, Michael (1972), *Painting and Experience in Fifteenth Century Italy*, Oxford University Press, paperback London.
Baxandall, Michael (1985), *Patterns of Intention*, Yale University Press, New Haven, CT.
Bazerman, Charles (1999), *The Languages of Edison's Light*, MIT Press, Cambridge, MA.
Beaudouin, Valérie and Julia Velkovska (1999), 'The Cyberians: an empirical study of sociality in a virtual community', in Kathy Buckner (ed.), *Proceedings of Esprit i3 Workshop on Ethnographic Studies in Real and Virtual Environments*, Queen Margaret College, Edinburgh, pp. 102–12.

Becker, Howard (1963), *Outsiders: Studies in the Sociology of Deviance*, Free Press, New York.

Béguin, Pascal (2007), 'Dialogisme et conception des systèmes de travail', *Psychologie de l'interaction*, **23–4**, 167–96.

Beltran, Alain and Patrice Carré (1991), *La Fée et la servante. La Société française face à l'électricité*, Belin, Paris.

Benjamin Walter (1989), 'Baudelaire', in *Paris, capitale du XIXe siècle*, Le Cerf, Paris, pp. 247–405.

Bertrand, Gisèle, Chantal de Gournay and Pierre-Alain Mercier (1988), 'Le programme global', *Réseaux*, **32**, 45–66.

Bijker, Wiebe (1992), 'The social construction of fluorescent lighting or how an artefact was invented in its diffusion stage', in Bijker and Law (eds), pp. 75–102.

Bijker Wiebe (1995), *Of Bicycles, Bakelites and Bulbs: Towards a Theory of Sociotechnical Change*, MIT Press, Cambridge, MA.

Bijker, Wiebe, Thomas Hughes and Trevor Pinch (eds) (1987), *The Social Construction of Technological System*, MIT Press, Cambridge, MA.

Bijker, Wiebe and John Law (eds) (1992), *Shaping Technology/Building Society: Studies on Sociotechnical Change*, MIT Press, Cambridge, MA.

Bijker, Wiebe and Trevor Pinch (2002), 'SCOT answers, other questions', *Technology and Culture*, **43** (2), 361–70.

Bimber, Bruce (1994), 'Three faces of technological determinism', in Roe Smith and Marx (eds), pp. 79–100.

Blaug, Mark (1963), 'A survey on the theory of process innovations', *Economica*, February, 13–32.

Bloch Marc (1935a), 'Avènement et conquêtes du moulin à eau', *Les Annales d'histoire économique et sociale*, **36**, 538–63, trans. English (1967), 'The advent and triumph of the watermill', in *Land and Work in Medieval Europe*, University of California Press, Berkeley, CA.

Bloch, Marc (1935b), 'Les Inventions médiévales', *Les Annales d'histoire économique et sociale*, **36**, republished in *Mélanges historiques* (1963), Sevpen, Paris, Vol. II, 822–8.

Bloch, Marc (1937), 'Que demander à l'histoire', republished in Vol. 1, pp. 3–15, *Mélanges Historiques* (1963), Sevpen, Paris.

Bloch Marc (1938), 'Technique et évolution sociale: réflexions d'un historien', *Europe*, republished in *Mélanges Historiques* (1963), Sevpen, Paris, Vol. II, 833–8.

Bloor, David (1976), *Knowledge and Social Imagery*, Routledge & Kegan Paul, London.

Boudon Raymond (1990), *L'Art de se persuader*, Fayard, Paris, trans. English (1994), *The Art of Self-persuasion: The Social Explanation of False Beliefs*, Polity, Cambridge.

Boullier, Dominique (1989), 'Du bon usage d'une critique du modèle diffusionniste', *Réseaux*, **36**, June, 31–51.

Boyer, Robert (ed.) (1991), *Les Figures de l'irréversibilité*, Éditions de l'École des Hautes Études en Sciences Sociales, Paris.

Braudel, Fernand (1979), *Civilisation matérielle, économie et capitalisme*, Vol. I: *Les Structures du quotidien*, Armand Colin, Paris, trans. English (1981), *Civilization and Capitalism, 15th–18th Century*, Vol. I: *The Structure of Everyday Life*, Harper & Row, New York.

Breton, Philippe (1992a), 'L'esprit et la matière, bref plaidoyer pour une sociologie amontiste des techniques', in Prades (ed.), pp. 45–52.

Breton, Philippe (1992b), *L'Utopie de la communication*, La Découverte, Paris.

Breton, Philippe (1993), *La Formation des valeurs et le champ de la sécurité informatique. Étude d'un corpus de textes de la littérature professionnelle*, CNET, mimeo.

Brown, Jerry (February 1978), Esquine cited in Theodore Roszack (1986), *The Cult of Information*, Lutterworth Press, Cambridge, MA, p. 65.

Brown, John Seely and Paul Duguid (1994), 'Borderline issues: social and material aspects of design', *Human–Computer Interaction*, **9**, 3–36.

Bruton, H.J. (1963), 'Contemporary theorizing on economic growth', cited by Blaug, pp. 287–97.

Bucciarelli, Louis (1994), *Designing Engineers*, MIT Press, Cambridge, MA.

Bud Frierman, Lisa (ed.) (1994), *Information Acumen: The Understanding and Use of Knowledge in Modern Business*, Routledge, London.

Callon, Michel (1980), 'Struggles and negotiations to define what is problematic and what is not: the sociologic of translation', in Knorr, Krohn and Whitley (eds), *The Process of Scientific Investigation*, D. Reide: Dordrecht, pp. 197–220.

Callon, Michel (1981), 'Pour une sociologie des controverses technologiques', *Fundamenta Scientae*, **2** (3/4), 381–99.

Callon, Michel (1986), 'Eléments pour une sociologie de la traduction', *L'Année sociologique*, PUF, Paris, pp. 169–208, trans. English 'Some elements of a sociology of translation', in Law (ed.), pp. 196–223.

Callon, Michel (1987), 'Society in the making: the study of technology as of tool for sociological analysis', in Bijker et al. (eds), pp. 83–103.

Callon, Michel (ed.) (1989a), *La Science et ses réseaux, genèse et circulation des faits scientifiques*, La Découverte, Paris.

Callon, Michel (1989b), 'L'agonie d'un laboratoire', in Callon (ed.) (1989a), pp. 173–214.
Callon, Michel and Bruno Latour (1981), 'Unscrewing the big Leviathan: how actors macro-structure reality and how sociologists help to do so', in Knorr-Cetina and Cicourel (eds), pp. 275–303.
Carey, James and John Quirk (1970), 'The mythos of the electronic revolution', *The Scholar*, **39** (2), Spring, 219–41.
Caron, François (1991), 'Histoire économique et dynamique des structures', *L'Année sociologique*, PUF, Paris.
Castoriadis, Cornelius (1992), 'Technique', *Encyclopedia Universalis*, Vol. 22, Paris, p. 126.
Castronova, Edward (2001), 'Virtual Worlds': a first-hand account of market and society on the Cyberian frontier', CESifo Working Papers 618, Munich.
Ceruzzi, Paul (1983), *Reckoners: the Prehistory of the Digital Computer 1935–1945*, Greenwood Press, Westport, CT.
Ceruzzi, Paul (1987), 'An unforeseen revolution: computers and expectations, 1935–1985', in Corn, (ed.), pp. 188–201.
Chanaron, Jean-Jacques and Jacques Perrin (1986), 'Science, technologie et modes d'organisation du travail', *Sociologie du travail*, no. 1, 23–40.
Chandler, Alfred (1990), *Scale and Scope: The Dynamics of Industrial Capitalism*, Harvard University Press, Cambridge, MA.
Chase, Stuart (1933), 'A vision in kilowatts', *Fortune*, April, p. 5.
Chevalier, Michel (1836), *Les Lettres sur l'Amérique du Nord*, Vol. II, Gosselin, Paris.
Clark, Kim and Steven Wheelwright (1992), 'Organizing and leading "heavyweight" development teams', *California Management Review*, Spring, 9–28.
Clarke, Adele (1990), 'A social worlds research adventure. The case of reproductive science', in Cozzens and Gieryn (eds), pp. 15–43.
Clarke, Adele (1991), 'Social worlds/arenas theory as organizational theory', in Maines (ed.), pp. 119–58.
Clayton Nick (2002), 'SCOT: does it answer?', *Technology and Culture*, **43** (2), 351–60.
Coleman, James, Elihu Katz and Herbert Menzel (1957), 'The diffusion of an innovation among physicians', *Sociometry*, December, 253–70.
Collins, Harry (1985), *Change in Order: Replication and Induction in Scientific Practice*, Sage, London.
Collins, Harry and Steven Yearley (1991), 'Epistemological chicken' in Pickering (ed.), pp. 369–89.
Constant, Edward (1973), 'A model for technological change applied to the turbojet revolution', *Technology and Culture*, **14** (4), 553–72.

Corn, Joseph (1983), *The Winged Gospel: America's Romance with Aviation 1900–1950*, Oxford University Press, Oxford and New York.
Corn, Joseph (ed.), (1987), *Imagining Tomorrow: History, Technology and the American Future*, MIT Press, Cambridge, MA.
Cortada, James (1993), *The Computer in the United States: From Laboratory to Market, 1930 to 1960*, Sharpe, Armonk, NY.
Coulon, Alain (1993), *L'Ethnométhodologie*, 'Que sais-je?', PUF, Paris.
Cowan, Ruth Schwartz (1983), *More Work for Mother: The Ironies of Household Technology from the Open Hearth to the Microwave*, Basic Books, New York.
Cowan, Ruth Schwartz (1987), 'The consumption junction: a proposal for research strategies in the sociology of technology', in Bijken et al. (eds), pp. 261–80.
Cowan, Robin (1990), 'Nuclear power reactors: a study in technology lock-in', *Journal of Economic History*, **50** (3), 541–66.
Cozzens, Susan and Thomas Gieryn (eds) (1990), *Theories of Science in Society*, Indiana University Press, Bloomington, IN.
Cresswell, Robert (1983), 'Transferts de techniques et chaînes opératoires', *Techniques et culture*, **2**, 143–59.
Cros, Charles (1970), *Étude sur les moyens de communication avec les planètes* (1869), in *Œuvres complètes*, Gallimard, 'La Pléiade', Paris, pp. 510–25.
Danto, Arthur (1965), *Analytical Philosophy of History*, Cambridge University Press, Cambridge.
Daumas, Maurice (ed.) (1962), *Histoire générale des techniques*, Vol. I, PUF, Paris.
Daumas, Maurice (1969), L'histoire des techniques: son objet, ses limites, ses méthodes', *Revue d'histoire des sciences et de leurs applications*, Paris, **22** (1), 5–32.
Daumas, Maurice (ed.) (1979), *Histoire générale des techniques*, Vol. V, PUF, Paris.
David, Paul (1971), 'The landscape and the machine: technical interrelatedness, land tenure and the mechanization of the corn harvest in Victorian Britain', in McCloskey (ed.), pp. 145–205.
David, Paul (1985), 'Clio and the economics of QWERTY', *American Economic Review*, **75** (2), May, 332–7.
David, Paul (1986), 'Understanding the economics of QWERTY: the necessity of history', in Parker (ed.), pp. 30–49.
David, Paul (1991), 'The hero and the herd in technological history: reflections on Thomas Edison and the battle of the systems', in Higonnet et al. (eds), pp. 72–119.

David, Paul (1992), 'Dépendance du chemin et prévisibilités dynamiques avec externalités de réseau localisées: un paradigme pour l'économie historique', in Foray and Freeman (eds), pp. 241–73.
David, Paul and Dominique Foray (1994), 'Percolation structures, Markov random fields and the economics of EDI standards diffusion', in Pogorel (ed.), pp. 135–70.
De Certeau, Michel (1980), *L'invention du quotidien*, Vol. I., *Arts de faire*, UGE-10/18, Paris, trans. English (1984), *The Practice of Everyday Life*, University of California Press, Berkeley, CA.
De Fleur, Melvin (1970), 'Mass communication and social change', (first published, 1966 *Social Forces*), in Tunstall (ed.), pp. 59–78.
De Fornel, Michel (1988), 'Contraintes systématiques et contraintes rituelles dans l'interaction visiophonique', *Réseaux*, **29**, 33–46.
De Fornel, Michel (1991), *Usages et pratiques du visiophone à Biarritz: objet technique, cadre interactionnel et sociabilité ordinaire*, CNET, mimeo, Paris.
De Fornel, Michel (1996), 'The interactional frame of videophonic exchange', *Réseaux, French Journal of Communication*, 4 (1), University of Lutton Press, Lutton, 47–72.
De L'Isle-Adam, Villiers (1986), *L'Ève future*, in *Œuvres complètes*, Vol. I, Gallimard, 'La Pléiade', Paris, trans. English (1981), *Eve of the Future Eden*, Coronado Press, Lawrence, KS.
De Michelis, Giorgio, Carla Simone and Kjeld Schmidt (eds) (1993), *Proceedings of the Third European Conference on Computer Supported Cooperative Work*, Kluwer Academic, Dordrecht.
De Oliviera Domingues, C. (1986), 'Technologie et crise', PhD thesis in economics, University of Paris-X-Nanterre.
Debray, Régis (1991), *Cours de médiologie générale*, Gallimard, Paris.
Denzin, Norman (ed.) (1978), *Studies in Symbolic Interaction*, JAI Press, Greenwich, CT.
Dixon, Robert (1980), 'Hybrid corn revisited', *Econometrica*, September, 1451–61.
Dockès, Pierre (1990), 'Formations et transferts des paradigmes sociotechniques', *Revue française d'économie*, **5** (4), Autumn, 29–82.
Dockès, Pierre and Bernard Rosier (1991), 'Histoire "raisonnée" et économie historique', *Revue économique*, **42** (2), March, 181–210.
Dosi, Giovanni (1982), 'Technological paradigms and technological trajectories', *Research Policy*, **11**, 147–62.
Dosi, Giovanni, Christopher Freeman, Richard Nelson, Gerald Silverberg and Luc Soete (eds) (1988), *Technical Change and Economic Theory*, Pinter, London.

Dosi, Giovanni and Luigi Orsenigo (1988), 'Coordination and transformation: an overview of structures, behaviours and change in evolutionary environments', in Dosi et al. (eds), pp. 13–37.

Douglas, Susan (1987), *Inventing American Broadcasting (1899–1922)*, Johns Hopkins University Press, Baltimore, MD.

Du Gay, Paul, Stuart Hall, Linda Janes, Hugh Mackay and Keith Negus (1997), *Doing Cultural Studies: The Story of the Sony Walkman*, Sage, London.

Dupuy, Jean-Pierre, François Eymard-Duvernay, Oliver Favereau, André Orléan, Robert Salais and Laurent Thévenot (1989), L'économie des conventions', *Revue Economique*, **40** (2), 141–6.

Durand, Claude (1978), *Le Travail enchaîné. Organisation du travail et domination sociale*, Le Seuil, Paris.

Eisenstein, Elizabeth L. (1993), *The Printing Revolution in Early Modern Europe*, Cambridge University Press, Cambridge, canto edition.

Ellul, Jacques (1977), *Le Système technicien*, Calmann-Lévy, Paris, trans. English (1980), *The Technological System*, Continuum, New York.

Ellul, Jacques (1990), *La Technique ou l'enjeu du siècle*, Economica, Paris, (1st edn 1954), trans. English (1964), *The Technological Society*, Knopf, New York.

Febvre, Lucien (1935), 'Réflexions sur l'histoire des techniques', *Annales d'histoire économique et sociale*, **36**, November, 531–5.

Fischer, Claude (1992a), 'Appels privés, significations individuelles. Histoire sociale du téléphone avant guerre aux États-Unis', *Réseaux*, **55**, 65–105.

Fischer, Claude (1992b), *America Calling: A Social History of the Telephone*, University of California Press, Berkeley, CA.

Flammarion, Camille (1911), *Mémoires biographiques et philosophiques d'un astronome*, E. Flammarion, Paris.

Fleck, James, Juliet Webster and Robin Williams (1990), 'Dynamics of information technology implementation', *Futures*, July–August, 618–40.

Flichy, Patrice (1991), 'The losers win. A comparative history of two innovations: videotex and the videodisc', in Jouët et al. (eds), pp. 73–86.

Flichy, Patrice (1995), *Dynamics of Modern Communication: The Shaping and Impact of New Communication Technologies*, Sage, London.

Flichy, Patrice (2007), *The Internet* Imaginaire, MIT Press, Cambridge, MA.

Foray, Dominique (1989), 'Les modèles de la compétition technologique. Une revue de la littérature', *Revue d'économie industrielle*, **48**, 2nd quarter, 16–34.

Foray, Dominique (1992), 'Choix des techniques, rendements croissants et processus historiques: la nouvelle économie du changement technique', in Prades (ed.), pp. 57–93.

Foray, Dominique and Christopher Freeman (eds) (1992), *Technologie et richesse des nations*, Economica, Paris.

Foray, Dominique and Christian Le Bas (1986), 'Diffusion de l'innovation dans l'industrie et fonction de recherche technique: dichotomie ou intégration', *Économie appliquée*, no. 3, 615–50.

France, Anatole (1890), 'Le café procope' in Charles Cros (ed.), *L'univers Illustré*, pp. 1220–21.

France, Anatole (1905 [1991]), *Sur la pierre blanche*, in *Œuvres*, Vol. III, Gallimard, 'La Pléiade', Paris, trans. English (1910), *The White Stone*, John Lane, London.

Freeman, Christopher (1982), *The Economics of Industrial Innovation*, France Pinter, London.

Freeman, Christopher (1985), 'Long waves of economic development', in Tom Forester (ed.), *The Information Technology Revolution*, Basil Blackwell, Oxford, pp. 602–16.

Freeman, Christopher (1986), 'Technologies nouvelles, cycles économiques longs et avenir de l'emploi', in Jean-Jacques Salomon and Geneviève Schméder (eds), *Les Enjeux de changement technologique*, Economica, Paris, pp. 91–108.

Freeman, Christopher (1988), 'A quoi tiennent la réussite ou l'échec des innovations dans l'industrie?', *Culture technique*, **18**, CRCT, Neuilly.

Freeman, Christopher, John Clark and Luc Soete (1982), *Unemployment and Technical Innovation: A Study of Long Waves and Economic Development*, Frances Pinter, London.

Freeman, Christopher and Carlota Perez (1988), 'Structural crises of adjustment business cycles and investment behaviour', in Dosi et al. (eds), pp. 38–66.

Freyssenet, Michel (1990a), *Les Formes sociales d'automatisation*, Cahiers du GIP Mutations Industrielles, Paris.

Freyssenet, Michel (1990b), *Les Techniques productives sont-elles prescriptives? L'exemple des systèmes experts en entreprise*, Cahiers du GIP Mutations Industrielles, Paris.

Freyssenet, Michel (1992), 'Processus et formes sociales d'automatisation. Le paradigme sociologique', *Sociologie du travail*, no. 4, 469–96.

Fridenson, Patrick (1988), 'Genèse de l'innovation: la 2 CV Citroën', *Revue française de gestion*, **70**, September–October, 35–44.

Friedmann, Georges (1966), *Sept Études sur l'homme et la technique*, Denoël/Gonthier, Paris.

Fujimura, Joan (1991), 'On methods, ontologies and representation in the sociology of science: where do we stand', in Maines (ed.), pp. 207–48.
Furet, François and Jacques Ozouf (1977), *Lire et écrire. L'alphabétisation des Français de Calvin à Jules Ferry*, Éditions de Minuit, Paris, Vol. I, trans. English (1982), *Reading and Writing: Literacy in France from Calvin to Jules Ferry*, Cambridge University Press, Cambridge and New York.
Gallie, D. (1978), *In Search of the New Working Class*, Cambridge University Press, Cambridge.
Garfinkel, Harold (1952), 'The perception of the other: a study in social order', unpublished PhD thesis (cited in Heritage, 1987).
Garfinkel, Harold, Michael Lynch and Eric Livingston (1981), 'The work of a discovering science construed with materials from the optically discovered pulsar', *Philosophy of the Social Sciences*, no. 11, 131–58.
Gaudin, Thierry (1984), *Pouvoir du rêve*, CRCT, Neuilly.
Gerson, Elihu and Susan Star (1986), 'Analysing due process in the work place', *ACM Transactions of Office Information Systems*, **4** (3), July, 257–70.
Giard, Luce (1990), *Présentation de la nouvelle édition des 'Arts de faire'*, Gallimard, Paris.
Giddens, Anthony and Jonathan Turner (eds) (1987), *Social Theory Today*, Stanford University Press, Stanford, CA.
Giedion, Siegfried (1983), *La Mécanisation au pouvoir*, Vol. I, Centre Pompidou/Centre de création industrielle, Denoël/Gonthier, Paris.
Gilfillan, S. Colum (1935), *Sociology of Invention*, Follett, Chicago.
Gille, Bertrand (1978), *Histoire des techniques*, Gallimard, 'La Pléiade', Paris.
Giraud, Alain (1993), 'L'Arlésienne du droit. Brève histoire du monopole de l'audiovisuel en France', *Réseaux*, **59**, CNET, Paris, 53–64.
Goffman, Erving (1974), *Frame Analysis: An Essay on the Organization of Experience*, Harper Colophon Books, New York.
Goffman, Erving (1981), *Form of Talk*, University of Pennsylvania Press, Philadelphia, PA.
Gouletquer, Pierre (1991), 'A propos de l'article Jouke S. Wigboldus, "Salt and crop production in the precolonial Central Sudan"', *Techniques et culture*, **17–18**, 1–35.
Granovetter, Mark (1978), 'Threshold models of collective behavior', *American Journal* of Sociology, **83** (6), 1420–43.
Gras, Alain (1993), *Grandeur et dépendance. Sociologie des macro-systèmes techniques*, PUF, Paris.
Griliches, Zvi (1957), 'Hybrid corn: an exploration in the economics of technological change', *Econometrica*, October, **25** (4), 501–22.

Guterl, Fred (1984), 'Design case history: Apple's Macintosh', *IEE Spectrum*, December, 34–5.
Hall, Stuart (1981), 'Encoding, decoding in television discourse', in Hall, Dorothy Hobson, Andrew Lowe and Paul Willis (eds), *Culture, Media, Language*, Hutchinson, London, pp. 128–38.
Haudricourt, André-Georges (1987), *La Technologie science humaine*, Éditions de la maison des sciences de l'homme, Paris.
Hawkes, Terence (1977), *Structuralism and Semiotics*, University of California Press, Berkeley, CA.
Hawkins, Richard (1993), *Public Standards and Private Networks: Some Implications of the Mobility Imperatives*, Science and Technology Policy Research Unit (SPRU), University of Sussex, Brighton.
Heilbroner, Robert L. (1994), 'Do machines make history?' (first published, 1967 *Technology and Culture*), in Roe Smith and Marx (eds), pp. 67–78.
Hennion, Antoine (1993), *La Passion musicale: une sociologie de la médiation*, Éditons Métailié, Paris.
Hennion, Antoine (1995), 'Mediation and social history of art', *Réseaux, The French Journal of Communication*, **3** (2), John Libbey, London, 233–62.
Heritage, John C. (1987), 'Ethnomethodology', in Giddens and Turner (eds), pp. 224–72.
Herskovits, Melville (1948), *Man and His Works: The Science of Cultural Anthropology*, A.A. Knopf, New York.
Hert, Philippe (1997), 'The dynamics of online interactions in a scholarly debate', *The Information Society*, **13** (4), 329–60.
Higonnet, Patrice, David Landes and Henry Rosovsky (eds) (1991), *Favorites of Fortune: Technology Growth and Economic Development since the Industrial Revolution*, Harvard University Press, Cambridge, MA.
Hiltz, Starr Roxane and Murray Turoff (1978), *The Network Nation: Human Communication via Computer*, Addison-Wesley, Reading, MA.
Hoddeson, Lilian (1981), 'The emergence of basic research in the Bell telephone system, 1875–1915', *Technology and Culture*, **22** (3), 512–44.
Hoffsaes, Colette (1978), 'L'informatique dans l'organisation: changement ou stabilité?', *Sociologie du travail*, no. 3, 280–309.
Hounshell, David A. (1975), 'Elisha Gray and the telephone: on the disadvantages of being an expert', *Technology and Culture*, **16** (2), 133–61.
Hounshell, David (1984), *From the American System to Mass Production*, Johns Hopkins University Press, Baltimore, MD.
Hughes, Thomas (1979), 'The electrification of America: the system builders', *Technology and Culture*, **20** (1), 124–61.

Hughes, Thomas (1983), *Networks of Power: Electrification in Western Society (1880–1930)*, Johns Hopkins University Press, Baltimore, MD.
Hughes Thomas (1987), 'Visions of electrification and social change', in *Un siècle d'électricité dans le monde*, Association pour l'histoire de l'électricité en France, PUF, Paris, pp. 327–40.
Hughes, Thomas (1994), 'Technological momentum', in Roe Smith and Marx (eds), pp. 101–13.
Innis, Harold (1951), *The Bias of Communication*, University of Toronto Press, Toronto.
Isambert, François-André (1985), 'Un "programme fort" en sociologie de la science?', *Revue française de sociologie*, **26** (3), 485–508.
Jennings, Tom (1998), 'Artist statement', available at www.wps.com/about-WPS/WPS/artist-statement.html, accessed 31 May, 2007.
Jouët, Josiane (1990), 'L'informatique "sans le savoir"', *Culture technique*, **21**, July, 216–22.
Jouët, Josiane (1991), 'A telematic community: the axiens', in Jouët et al. (eds) pp. 181–202.
Jouët Josiane (1994), 'Communication and mediation', *Réseaux. The French Journal of Communication*, 2 (1), John Libbey, London, 73–90.
Jouët, Josiane (2000), 'Retour critique sur la sociologie des usages', *Réseaux*, **100**, 487–522.
Jouët, Josiane, Patrice Flichy and Paul Beaud (eds) (1991), *European Telematics: The Emerging Economy of Words*, North-Holland, Amsterdam.
Jouët, Josiane and Dominique Pasquier (1999), 'Youth and screen culture: national survey of 6–17 year olds', *Réseaux, The French Journal of Communication*, **7** (1), University of Lutton Press, Lutton, 29–58.
Katz, Elihu (1971), 'The social itinerary of technical change: two studies of the diffusion of innovation', in Schramm and Roberts (eds), pp. 760–97.
Katz, Elihu, Hadasah Haas and Michael Gurevitch (1973), 'On the use of the mass media for important things', *American Sociological Review*, **38** (2), 164–81.
Kline, Ronald (2003), 'Resisting consumer technology in rural America: the telephone and electrification', in Oudshoorn and Pinch (eds), pp. 51–66.
Kline, Ronald and Trevor Pinch (1996), 'Users as agents of technological change: the social construction of the automobile in the rural United States', *Technology and Culture*, **37** (4), 763–95.
Kling, Rob (1996), *Computerization and Controversy: Value Conflicts and Social Choices*, Academic Press, San Diego.

Knight, Kenneth (1967), 'A descriptive model of the intra-firm innovation process', *Journal of Business*, October, **40** (4), 478–96.

Knorr-Cetina, Karin, Roger Krohn and Richard Whitley (eds) (1980), *The Social Process of Scientific Investigation: Sociology of the Sciences Yearbook*, Vol. 4, D. Reidel Dordrecht.

Knorr-Cetina, Karin and Aaron V. Cicourel (eds) (1981), *Advances in Social Theory and Methodology: Toward an Integration of Micro and Macro-Sociologies*, Routledge & Kegan Paul, Boston, MA.

Knorr-Cetina Karin, Michael Mulkay (eds) (1983), *Science Observed: Perspectives in the Social Studies of Science*, Sage, London.

Kuhn, Thomas (1962), *The Structure of Scientific Revolutions*, University of Chicago Press, Chicago.

Lacaze, Dominique (1979), 'Lectures croisées de Jules Verne et de Robida', in Raymond and Vierne (eds), pp. 76–90.

Landes, David (1991), 'On technology and growth', in Higonnet et al. (eds), pp. 1–29.

Latour, Bruno (1985), 'Les vues de l'esprit. Visualisation et connaissance scientifique', *Culture technique*, **14**, 5–29.

Latour, Bruno (1987), *Science in Action*, Harvard University Press, Cambridge, MA.

Latour, Bruno (1989a), 'Joliot: l'histoire et la physique mêlées', in Serres (ed.), pp. 493–513.

Latour, Bruno (1989b), 'Pasteur et Pouchet: hétérogénèse de l'histoire des sciences', in Serres (ed.), pp. 423–45.

Latour, Bruno (1992a), 'Il faut qu'une porte soit ouverte ou fermée … Petite philosophie des techniques', in Prades (ed.), pp. 27–40.

Latour, Bruno (1992b), *Aramis ou l'amour des techniques*, La Découverte, Paris, trans. English (1996), *Aramis, or, The Love of Technology*, Harvard University Press, Cambridge, MA.

Latour, Bruno (1996), *Petites leçons de sociologie des sciences*, La Découverte, Paris.

Latour, Bruno (2005) *Reassembling the Social: An Introduction to Actor-Network-Theory*, Oxford University Press, Oxford.

Latour, Bruno, Philippe Mauguin and Geneviève Teil (1992), 'A note on socio-technical graphs', *Social Studies of Science*, **22**, 33–57.

Latour, Bruno and Steve Woolgar (1979), *Laboratory Life: The Construction of Scientific Facts*, Princeton University Press, Princeton, NJ.

Laughlin, C.D. (1983), 'Les artefacts de la connaissance. Une perspective biogénétique structurale du symbole et de la technologie', *Anthropologie et Sociétés*, **13** (2), 9–29.

Law, John (ed.) (1986), *Power, Action and Belief: A New Sociology of Knowledge?* Routledge, London.

Law, John (ed.) (1991), *A Sociology of Monsters*, Routledge, London.

Lazarsfeld, Paul, Bernard Berelson and Hazel Gaudet (1944), *The People's Choice*, Columbia University Press, New York.

Le Bas, Christian (1982), *Économie des innovations techniques*, Économica, Paris.

Lefebvre des Noëttes, Richard (1931), *L' Attelage et le Cheval de selle à travers les âges*, Picard, Paris.

Leroi-Gourhan, André (1936), *La Civilisation du renne*, Gallimard, Paris.

Leroi-Gourhan, André (1950a) *Évolution et technique*, I, *L'Homme et la matière*, Albin Michel, Paris (1st edn 1943).

Leroi-Gourhan, André (1950b), *Évolution et technique*, II, *Milieu et techniques*, Albin Michel, Paris.

Leroi-Gourhan, André (1964), *Le Geste et la parole*, I, *Technique et langage*, Albin Michel, Paris.

Leroi-Gourhan, André (1965), *Le Geste et la parole*, II, *La Mémoire et les rythmes*, Albin Michel, Paris, trans. English (1993), *Gesture and Speech*, MIT Press, Cambridge, MA.

Leroi-Gourhan, André (1968), 'L'expérience ethnologique', in Poirier (éd.), pp. 1816–25.

Lévy, Pierre (1987), *La Machine univers*, La Découverte, Paris.

Levy, Steven (1985), *Hackers: Heroes of the Computer Revolution*, Dell Book, New York.

Licklider, Joseph and Robert Taylor (1968), 'The computer as a communication device', *Science and Technology*, April pp. 21–31, re-published (1990) in *In Memoriam: J.C.R. Licklider 1915–1990*, Digital Systems Research Center, Palo Alto, CA, pp. 21–41.

Lievrouw, Leah (2006), 'New media design and development: diffusion of innovations v social shaping of technology', in Lievrouw and Sonia Livingstone (eds), *The Handbook of New Media*, Sage, London, pp. 246–65.

Linhart, Danièle (1991), *Le Torticolis de l'autruche. L'éternelle modernisation des entreprises françaises*, Le Seuil, Paris.

Livet, Pierre and Laurent Thévenot (1991), *L'Action collective*, colloque 'L'Économie des conventions', March, Paris, 22–9.

Livingstone, Sonia (2003), 'The changing nature of audiences: from the mass audience to the interactive media user', LSE Research Online, London http://eprints.lse.ac.uk/archive/00000417, accessed 31 May 2007.

Lorenz, Konrad (1976), *L'Envers du miroir*, Flammarion, Paris, trans. English (1977), *Behind the Mirror: A Search for a Natural History of Human Knowledge*, Methuen, London.

Lynch, Michael (1993), *Scientific Practice and Ordinary Action: Ethnomethodology and Social Studies of Science*, Cambridge University Press, Cambridge and New York.

Lynch, Michael, Eric Livingston and Harold Garfinkel (1983), 'Temporal order in laboratory work', in Knorr-Cetina and Mulkay (eds), pp. 205–38.

Mackay, Hugh, Chris Carne, Paul Beynon-Davies and Doug Tudhope (2000), 'Reconfiguring the User: Using rapid application development', *Social Studies of Science*, **30** (5), 737–57.

MacKenzie, Donald and Judy Wajcman (eds) (1985), *The Social Shaping of Technology*, Open University Press, Milton Keynes.

Maines, David (ed.) (1991), *Social Organization and Social Process: Essays in Honor of Anselm Strauss*, Aldine de Gruyter, New York.

Malerba, Franco (1985), 'Demand structure and technology change: the case of the European semi-conductor industry', *Research Policy*, **14**, 283–97.

Mallet, Serge (1969), *La Nouvelle Classe ouvrière* (1st edn 1963), Le Seuil, Paris, trans. English (1975), *The New Working Class*, Bertrand Russell Peace Foundation for Spokesman Books, Nottingham.

Mangematin, Vincent (1993), 'Compétition technologique: les coulisses de la mise sur le marché', *Annales des mines*, June 4–16.

Mansfield, Edwin (1961), 'The technical change and the rate of imitation', *Econometrica*, October, **29** (4), 741–66.

Marshall, Alfred (1919), *Industry and Trade*, Macmillan, London.

Marti, Bernard (1994), 'La normalisation du vidéotex', in Feneyrol and Guérard (eds), *Innovation et Recherche Télécommunications*, Eyrolles, Paris, pp. 211–47.

Martin, Michèle (1991), *Hello, Central? Gender, Technology, and Culture in the Formation of Telephone Systems*, McGill-Queen's University Press, Montreal.

Marvin, Carolyn (1987), 'Dazzling the multitude: imagining the electric light as a communications medium', in Corn (ed.), pp. 202–17.

Marvin, Carolyn (1988), *When Old Technologies were New: Thinking about Electric Communication in the Late Nineteenth Century*, Oxford University Press, Oxford and New York.

Marx, Karl (1955), *The Poverty of Philosophy*, Progress Publishers, Moscow.

Marx, Leo (1964), *The Machine in the Garden Technology and the Pastoral Idea in America*, Oxford University Press, Oxford and New York.

Maunoury, Jean-Louis (1968), *La Genèse des innovations*, PUF, Paris.

Maunoury, Jean-Louis (1972), *Économie du Savoir*, Armand Colin, Paris.

Mauss, Marcel (1967), *Manuel d'ethnographie*, 'Petite bibliothèque Payot', Paris, pp. 30–32.
Mauss, Marcel (1985 [1950]), *Sociologie et Anthropologie*, PUF, Paris.
McCloskey, D. (ed.) (1971), *Essays on a Mature Economy: Britain after 1840*, Methuen, London.
McCloskey, D.M. (1976), 'Does the past have useful economics?', *Journal of Economic Literature*, June, 434–51.
McGuire, J.E. and T. Melia (1989), 'Some cautionary structures on the writing of the rhetoric of science', *Rhetorica*, 7, 87–99.
McLuhan, Marshall (1964), *Understanding Media: The Extensions of Man*, Routledge & Kegan Paul, London.
McLuhan, Marshall (1967), *The Gutenberg Galaxy: The Making of Typographic Man*, Routledge & Kegan Paul, London.
Mead, George H. (1932), 'The physical thing', in George Mead, *The Philosophy of the Present*, Open Court Publishing Chicago.
Méadel, Cécile (1990), 'De la formation des comportements et des goûts. Une histoire des sondages à la télévision, dans les années cinquante', *Réseaux*, **39**, CNET, Paris, 37–55.
Mendelbaum, Maurice (1977), *The Anatomy of Historical Knowledge*, Johns Hopkins University Press, Baltimore, MD (cited in Ricoeur, 1991, p. 349).
Mensch, Gerhard (1979), *Stalemate in Technology: Innovations Overcome the Depression*, Ballinger, New York.
Mensch, Gerhard (1988), 'La technique en crise', *Culture technique*, **18**, CRCT, Neuilly.
Midler, Christophe (1993), *L'auto qui n'existait pas. Management des projets et transformation de l'entreprise*, InterÉditions, Paris.
Misa, Thomas J. (1994), 'Retrieving sociotechnical change from technological determinism', in Roe Smith and Marx (eds), pp. 115–41.
Monjardet, Dominique (1980), 'Organisation, technologie, et marché de l'entreprise industrielle', *Sociologie du travail*, no. 1, 76–96.
Morley, David (1993), 'Active audience theory: pendulum and pitfalls', *Journal of Communication*, **43** (4), 13–19.
Mowery, David and Nathan Rosenberg (1979), 'The influence of market demand upon innovation: a critical review of some recent empirical studies', *Research Policy*, **8**, 102–53.
Mueller, Milton (1997), *Universal Service: Competition, Interconnection, and Monopoly in the Making of the American Telephone System*, MIT Press, Cambridge, MA.
Mumford, Lewis (1934), *Technics and Civilization*, Harcourt, Brace, New York.

Mumford, Lewis (1967), *The Myth of the Machine*, Harcourt, Brace, New York.
Naville, Pierre (1960), 'Vers l'automatisme social', *Revue française de sociologie*, February **24** (3), 275–85.
Nelson, Richard (1964), 'Aggregate production functions and medium range growth projections', *American Economic Review*, September, **54** (5), 575–606.
Nelson, Richard and Sidney Winter (1982), *An Evolutionary Theory of Economic Change*, Harvard University Press, Cambridge, MA.
Noble, David (1984), *Forces of Production: A Social History of Industrial Automation*, Alfred Knopf, New York.
Noble, David (1985), 'Social choice in machine design: the case of automatically controlled machine tools', in MacKenzie and Wajcman (eds), pp. 109–24.
Nouveau vocabulaire de la langue française (1837), *Nouveau vocabulaire de la langue française*, Lebigre frères, Paris.
Odasz, Frank (1995), 'Issues in the development of community cooperative networks', in Brian Kahn and James Keller (eds), *Public Access to the Internet*, MIT Press, Cambridge, MA.
Ogburn, William and Dorothy Thomas (1922), 'Are inventions inevitable? A note on social evolution', *Political Science Quarterly*, **37** (1), 83–98.
Orlikowski, Wanda (1992), 'The duality of technology: rethinking the concept of technology in organizations', *Organization Science*, **3** (3), 398–427.
Orlikowski, Wanda (2000), 'Using technology and constituting structures: a practice lens for studying technology in organizations', *Organization Science*, **11** (4), 404–28.
Orlikowski, Wanda and Debra Gash (1994), 'Technological frames: making sense of information technology in organizations', *ACM Transactions on Information Systems*, **12** (2), 174–207.
Orlikowski, Wanda, JoAnne Yates, Kazuo Okamura and Masayo Fujimoto (1995), 'Shaping electronic communication: the metastructuring of technology in the context of use' *Organization Science*, **6** (4), 423–44.
Ortsmann, Oscar (1978), *Changer le travail, les expériences, les méthodes, les conditions de l'expérimentation sociale*, Dunod, Paris.
Oudshoorn, Nelly and Trevor Pinch (eds) (2003), *How Users Matter: The Co-Construction of Users and Technologies*, MIT Press, Cambridge, MA.
Pacey, Arnold (1983), *The Culture of Technology*, MIT Press, Cambridge, MA.
Parker, W.N. (ed.) (1986), *Economic History and the Modern Economist*, Basil Blackwell, Oxford.

Perriault, Jacques (1989), *La Logique de l'usage. Essai sur les machines à communiquer*, Flammarion, Paris.
Perrin, Jacques (1991), 'Méthodologie d'analyse des systèmes techniques', in Boyer (ed.), pp. 151–71.
Perrot, Michelle (1979), 'Les problèmes de main-d'œuvre industrielle', in Daumas (ed.), pp. 477–509.
Pestre, Dominique (ed.) (1992), *L' Etude sociale des sciences. Bilan des années 1970 et 1980 et conséquences pour le travail historique*, CHRST, Cité des Sciences et de L'industrie, Paris.
Pharo, Patrick and Louis Quéré (eds) (1990), 'Les formes de l'action', *Raisons pratiques*, **1**.
Pickering, Andrew (ed.) (1991), *Science as Practice and Culture*, University of Chicago Press, Chicago.
Pinch, Trevor (2001), 'Why you go to a piano store to buy a synthesizer: path dependence and the social construction of technology', in Raghu Garud and Peter Karnøe (eds), *Path Dependence and Creation*, Lawrence Erlbaum Associates, Mahwah, NJ pp. 381–99
Pinch, Trevor and Wiebe Bijker (1987), 'The social construction of facts and artefacts: or how the sociology of science and the sociology of technology might benefit each other', in Bijker et al. (eds), pp. 17–50.
Pogorel, G. (ed.) (1994), *Global Telecommunication Strategies and Technological Change*, Elsevier, Amsterdam.
Poirier, Jean (ed.) (1968), *Ethnologie générale*, Gallimard, 'La Pléiade', Paris.
Prades, Jacques (ed.) (1992), *La Technoscience. Les fractures du discours*, L'Harmattan, Paris.
Preece, William (1898), *Review of Reviews*, December, p. 715, cited by Susan Douglas, 'Amateur operators and American broadcasting: shaping the future of radio', in Corn (ed.) (1987), p. 35.
Proulx, Serge (ed.) (1988), *Vivre avec l'ordinateur, les usagers de la micro-informatique*, G. Vermette, Quebec.
Quéré, Louis (1989), 'Les Boîtes noires de Bruno Latour ou le lien social dans la machine', *Réseaux*, **36**, June, 95–117.
Quéré, Louis (1992), 'Espace public et communication, remarques sur l'hybridation des machines et des valeurs', in Pierre Chambat (ed.), *Communication et lien social*, Descartes, Paris, pp. 29–49.
Rallet, Alain (1993), 'La théorie des conventions chez les économistes', *Réseaux*, **62**, 43–61.
Raymond, François and Simone Vierne (eds) (1979), *Jules Verne et les sciences humaines*, 10/18, UGE, Paris.
Real, Bernard (1990), *La Puce et le chômage. Essai sur la relation entre le progrès technique, la croissance et l'emploi*, Le Seuil, Paris.

Reynolds, Barrie (1983), 'The relevance of material culture to anthropology', *Journal of the Anthropological Society of Oxford*, **14** (1), 17–43.

Rheingold, Howard (1994), *The Virtual Community: Homesteading on the Electronic Frontier*, Harper Collins, New York.

Rice, Ronald and Everett Rogers (1980), 'Reinvention in the innovation process', *Science Communication*, **1** (4), 499–514.

Ricoeur, Paul (1977), *La Sémantique de l'action*, in Tiffeneau (ed.), pp. 21–63.

Ricoeur, Paul (1991), *Temps et récit*, I., *L'Intrigue et le récit historique*, Seuil, coll. 'Points', Paris, trans. English (1988), *Time and Narrative*, University of Chicago Press, Chicago.

Rip, Arie, Thomas Misa and Johan Schot (eds) (1995), *Managing Technology in Society*, Pinter, London.

Robbins, Lionel (1932), *An Essay on the Nature and Significance of Economic Science*, Macmillan, London.

Robinson, Mike (1993), 'Design for unanticipated use ...', in de Michelis et al. (eds), pp. 187–202.

Roe Smith, Merritt and Leo Marx (eds) (1994), *Does Technology Drive History? The Dilemma of Technological Determinism*, MIT Press, Cambridge, MA.

Rogers, Everett (1983), *Diffusion of Innovations* (3rd edn), Free Press, New York.

Rogers, Everett (1986), *Communication Technology: The New Media in Society*, Free Press, New York.

Rogers, Everett and Judith Larsen (1984), *Silicon Valley Fever: Growth of High Technology Culture*, Basic Books, New York.

Rosenberg, Nathan (ed.) (1982a), *Inside the Black Box: Technology and Economics*, Cambridge University Press, Cambridge.

Rosenberg, Nathan (1982b), 'How exogenous is science?', in Rosenberg (ed.), pp. 141–59.

Rosenberg, Nathan (1982c), 'Learning by using', in Rosenberg (ed.), pp. 120–40.

Rosenberg, Nathan (1982d), 'Marx as a student of technology', in Rosenberg (ed.), pp. 34–53.

Ryan, Bryce and Neal Gross (1943), 'The diffusion of hybrid seed corn in two Iowa Communities', *Rural Sociology*, **8**, 15–24.

Sadoul, Jacques (1984), *Histoire de la science-fiction moderne (1911–1984)*, Robert Laffont, Paris.

Sahal, Devendra (1985), 'Technology guide-posts and innovation avenues', *Research Policy*, **14** (2), 61–82.

Schmookler, Jacob (1966), *Invention and Economic Growth*, Harvard University Press, Cambridge, MA.

Schramm, Wilbur and Donald Roberts (eds) (1971), *The Process and Effects of Mass Communication* (first published 1954), University of Illinois Press, Urbana.

Schumpeter, Joseph (1939), *Business Cycles*, McGraw Hill, New York.

Schütz Alfred (1962), 'Common sense and scientific interpretations of human action', in Alfred Schütz *Collected Papers*, Vol. 1, Martinus Nijhoff, The Hague, (cited in Heritage, 1987, pp. 3–47).

Segal, Howard (1985), *Technological Utopianism in American Culture*, University of Chicago Press, Chicago.

Segrestin, Denis (2004), *Les Chantiers du manager*, Armand Colin, Paris.

Serres, Michel (1980), *Hermes V. Le Passage du Nord-Ouest*, Éditions de Minuit, Paris.

Serres, Michel (ed.) (1989), *Éléments d'histoire des sciences*, Bordas, Paris, trans. English (1995), *A History of Scientific Thought: Elements of a History of Science*, Blackwell, Oxford.

Shapin, Steve and Simon Schaffer (1985), *Leviathan and the Air-pump: Hobbes, Boyle, and the Experimental Life*, Princeton University Press, Princeton, NJ.

Shapiro, Carl and Hal Varian (1999), *Information Rules: A Strategic Guide to the Network Economy* Harvard Business School Press, Boston, MA.

Shibutani Tamotsu (1955), 'Reference groups as perspectives', *American Journal of Sociology*, **60** (6), 562–9.

Silverstone, Roger and Eric Hirsch (eds) (1994), *Consuming Technologies: Media and Information in Domestic Spaces*, Routledge, London.

Simondon, Gilbert (1989), *Du mode d'existence des objets techniques*, Aubier, Paris.

Singer, Charles, E.J. Holmgard and A.R. Hall (1954–84), *A History of Technology*, 5 vols, Clarendon, Oxford.

Solow, Robert (1957), 'Technical change and the aggregate production function', *Review of Economics and Statistics*, August, **39**, 312–20.

Star, Susan Leigh and James Griesemer (1989), 'Institutional ecology, translations and boundary objects: amateurs and professionals in Berkeley's Museum of Vertebrate Zoology (1907–1939)', *Social Studies of Sciences*, **19**, Sage, London, pp. 387–420.

Strauss, Anselm (1969), *Mirrors and Masks: The Search for Identity*, San Francisco Sociology Press.

Strauss, Anselm (1978), 'A social worlds perspective', in Denzin (ed.), pp. 119–28.

Suchman, Lucy (1987), *Plans and Situated Actions: The Problem of Human–machine Communication*, Cambridge University Press, Cambridge and New York.

Suchman, Lucy (1990), 'Plans d'action. Problèmes de représentation de la pratique en sciences cognitives', in Pharo and Quéré (eds), pp. 149–70.
Thévenot, Laurent (1993), 'A quoi convient la théorie des conventions?', *Réseaux*, **62**, CNET, Paris, 137–42.
Thuillier, Pierre (1987), 'La science existe-t-elle? Le cas Pasteur', *La Recherche*, April, **187**, 506–11.
Tiffeneau, Dorian (ed.) (1977), *La Sémantique de l'action*, CNRS, Paris.
Tunstall, Jeremy (ed.) (1970), *Media Sociology: A Reader*, Constable, London.
Turkle, Sherry (1982), 'The subjective computer: a study in the psychology of personal computation', *Social Studies of Science*, **12** (2), 173–205.
Usher, Abbott P. (1954), *A History of Mechanical Inventions* (1st edn, 1929), Harvard University Press, Cambridge, MA.
Utterback, James (1974), 'Innovation in industry and the diffusion of technology', *Science*, February, **183**, 595–674.
Valente, Thomas W. (1995), *Network Models of the Diffusion of Innovations*, Hampton Press Cresskill, NJ.
Van Lente, Harro and Arie Rip (1998), 'The rise of membrane technology: from rhetorics to social reality', *Social Studies of Science*, **28** (2), 221–54.
Velkovska, Julia (2002), 'The anonymous intimacy of electronic conversations in *Webchats*', *Sociologie du travail*, **44** (2), 193–213.
Vercruysse, Jean-Pierre and Pascal Verhoest (1994), 'The role of the state in the Belgian telecommunications sector in the 19th century', *Réseaux, The French Journal of Communications*, **2** (1), John Libbey, London, 91–112.
Verne, Jules (1888 [1994]), *Le Château des Carpathes*, Hachette, 'Livre de Poche', Paris, trans. English (1890), *The Castle of the Carpathians*, S. Low Marston, London.
Veyne, Paul (1971), *Comment on écrit l'histoire*, Le Seuil, Paris, trans. English (1984), *Writing History: Essay on Epistemology*, Wesleyan University Press, Middletown, CT.
Von Hayek, Friedrich (1953), *Scientisme et sciences sociales*, Plon, Paris.
Von Hippel, Eric (1988), *The Sources of Innovation*, Oxford University Press, Oxford and New York.
Von Hippel, Eric (2002), 'Horizontal innovation networks: by and for users', Working Paper 4366–02, MIT Sloan School of Management, Cambridge, MA.
Von Hippel, Eric (2005), *Democratizing Innovation*, MIT Press, Cambridge, MA.

Walras, Léon (1877), *Eléments d'économie politique pure (ou théorie de la richesse nationale)*, Paris (cited in Le Bas, 1982 p. 25), trans. English (1984), *Elements of Pure Economics*, Orion Editions, Philadelphia, PA.

Weber, Max (1965), 'Études critiques de logique des sciences', in *Essais sur la théorie de sciences*, Plon, Paris, 215–324.

White, Lynn (1962), *Medieval Technology and Social Change*, Clarendon Press, Oxford.

Williams, Raymond (1974), *Television, Technology and Cultural Form*, Fontana/Collins, London.

Williams, Robin and David Edge (1996), 'The social shaping of technology', *Research Policy*, **25** (6), 865–99.

Williams, Rosalind (1990), *Notes on the Underground: An Essay on Technology, Society and Imagination*, MIT Press, Cambridge, MA.

Winner, Langdon (1993), 'Upon opening the black box and finding it empty: social constructivism and the philosophy of technology', *Science Technology and Human Values*, **18** (3), Summer, 362–78.

Woolgar, Steve (1991), 'Configuring the user: the case of usability trials', in Law (ed.), pp. 57–99.

Wyatt, Sally (2003), 'Non-users also matter: the construction of users and non-users of the Internet', in Oudshoorn and Pinch (eds), pp. 67–79.

Yates, JoAnne (1989), *Control through Communication: The Rise of System in American Management*, Johns Hopkins University Press, Baltimore, MD.

Yates, JoAnne (1993), 'Co-evolution of information-processing technology and use: interaction between the life insurance and tabulating industries', *Business History Review*, **67** (1), 1–51.

Yates, JoAnne (1994), 'Evolving information use in firms, 1850–1920: ideology and information techniques and technologies', in Bud Frierman (ed.), 26–50.

Young, Jeffrey (1988), *Steve Jobs: The journey Is the Reward*, Scott, Foresman, Glenview, IL.

Zitt, Michel (1987), 'Filiations techniques et genèse de l'innovation', *Techniques et culture*, **10**, Editions de la Maison des sciences de l'homme, Paris, 13–43.

Zola, Émile (1901), *Le travail*, E. Fasquelle, Paris, trans. English (1901), *Labor*, Harper & Brothers, London.

Index

Abbate, Janet 138, 153
Abernathy, William 115, 161
AC/DC example 107–8
actants 58–9
action, definitions 75
action together 94
Actor Network Theory (ANT) 60, 62–5
actors, in Latour and Callon networks 58–60
Advanced Research Projects Agency (ARPA) 138
Aitken, Hugh 65, 147
Akrich, Madeleine 55, 59, 61, 66–7, 71–2
Alsène, Eric 37
Altair 136
Alter, Norbert 31, 33
alternative frames of use 84–5
Amblard, Henri 62–3
Ancelin, Claire 162
Apple 136–7
appropriation 68
arenas 78
Arnal, Nicole 93
Aron, Raymond 100, 109
Arpanet 138, 153
Arrow, Kenneth 8, 119
Arthur, Brian 102–6
ATT triode example 64–5
automatic metro (Aramis) example 59, 61

Bachelard, Gaston 50
Baczko, Bronislaw 125
Ballé, Catherine 33
Barbichon, Georges 31
Barère, Bertrand 149
Barras, Richard 22, 118
Baszanger, Isabelle 77
Battelle, John 161

Baudelaire, Charles 132
Baxandall, Michael 29, 91, 121, 154
Bazerman, Charles 134
BBS (bulletin board systems) 140
Beaudouin, Valérie 69
Becker, Howard 77
Béguin, Pascal 79
Bell, Alexander Graham 146
Bellamy, Edward 126
Beltran, Alain 143
Benjamin, Walter 132
Berliner, Emile 151
Bertrand, Gisèle 94
beta testers 94
bicycle examples 50–52
Bijker, Wiebe 49–52, 71–2, 80
Bimber, Bruce 25
Blaug, Mark 5, 7
Bloch, Marc 24, 119
Bloor, David 48
Boudon, Raymond 72
Boullier, Dominique 18
boundary frames 81
boundary-object phase 160–61
boundary objects 78–9
Bourdieu, Pierre 56
Boyer, Robert 119
Braudel, Fernand 24
Breton, Philippe 121–3
bridge construction example 91–2
Brown, Jerry 137
Bucciarelli, Louis 79, 148, 162
bulletin board systems (BBS) 140
Busson, Alain 93

Callon, Michel 54–7, 59–61
Carey, James 132
Caron, François 150
Carré, Patrice 143
Castoriadis, Cornelius 17–18, 39, 159
Castronova, Edward 94

189

catchall-object phase 158–9
causal imputation 100
Ceruzzi, Paul 123
Chanaron, Jean-Jacques 37–8
Chandler, Alfred 25
Chappe, Claude 134, 151, 154
Chase, Stuart 132
Chevalier, Michel 131, 144
cinema example 89
Citroën 2CV example 146–7
Civilisation matérielle, Èconomie et capitalisme (Braudel) 24
Clark, Kim 79, 115
Clarke, Adele 77–8
Clayton, Nick 72
coal mines examples 31–2
Coleman, James 11
Collins, Harry 48
common artifacts 85–6
common worlds 79
communications technology, effects 26–30
Community Memory 135–6
computer development example 121–3
computers *see* ICT studies
concretization 111–12
concurrent history phase 157–8
Constant, Edward 21, 113
consumer innovations 6–7
consumption junction 65
conventional objects 94
conversion of ICT objects 68
Corn, Joseph 130
Cortada, James 90
Coulon, Alain 96
Cowan, Robin 104, 106
Cowan, Ruth Schwartz 65, 68
Cresswell, Robert 46
critical mass 13
Cros, Charles 128–9
cultural focus 45–6
cultural technology 43–8

Danto, Arthur 99
Daumas, Maurice 14
David, Paul 105–8, 148
De Certeau, Michel 87–8, 92
De Fleur, Melvin 12, 26
De Forest, Lee 153
De Fornel, Michel 69, 85–6, 95
De Lamartine, Alphonse 125
De Oliviera Domingues, Cristiano 117
Debray, Régis 28
decoding 66–7
demand–pull theories 19–23
design engineers' culture 36–7
determinism
 historians' views 23–6
 see also Ellul, Jacques
detribalization 27
diffusion
 economic models 7–9
 sociological models 9–13
Dixon, Robert 7
Dockès, Pierre 107, 118–19, 162
Dolbear, Amos 129–30
domestication of ICT objects 68
Dosi, Giovanni 20, 110–11, 113–14
Douglas, Susan 130, 134
Du Gay, Paul 69
Du Mode d'existence des objets techniques (Simondon) 111
Dupuy, Jean-Pierre 94
Durand, Claude 33

Edge, David 49
Eisenstein, Elizabeth L. 29–30
electric vehicle example 56, 60
electricity *imaginaire* 131–3
Ellul, Jacques 15–17
entrepreneurs, definitions 77
ethnomethodology 73–6
Eve future (Villiers de L'Isle-Adam) 127–8
evolution of technical object 8
evolutionary economics 109–12
excluded users 69
expelled users 69
expert systems examples 36

Febvre, Lucien 13–15
Felsenstein, Lee 135–6
Fessenden, Reginald 153
Figuier, Louis 130
Fischer, Claude 96, 103, 151
Flammarion, Camille 129, 143
Fleck, James 35–6
Flichy, Patrice 12, 69, 72, 89–90, 96, 134, 140, 143, 149–51, 154, 159

Foray, Dominique 9, 18, 102–3, 105, 107–9, 148
Ford/GM example 89
Forth rail bridge example 91–2
Fragments d'histoire future (Tarde) 128
frames of functioning 82
  see also *imaginaire, technological imaginaire* and frame of functioning
frames of reference 80–87, 123–4
frames of use 82–5
France, Anatole 128–9
Freeman, Christopher 22–3, 90, 114–15, 117–19
Freyssenet, Michel 36–7
Fridenson, Patrick 146
Friedmann, Georges 32
Fujimura, Joan 76
Fulchiron, Jean-Claude 150
Furet, François 28–9

Gallie, D. 33
Garfinkel, Harold 74–5, 96
Gash, Debra 70
Gaston Lagaffe 56
Gates, Bill 137
Gaudin, Thierry 120, 125, 146
Geddes, Patrick 131
Gernsback, Hugo 129
Gerson, Elihu 76
Giard, Luce 92
Giedion, Siegfried 37
Gilfillan, S. Colum 25
Gille, Bertrand 14–15, 104–5, 107
Giraud, Alain 152
Goffman, Erving 80, 85
Google 161
Gouletquer, Pierre 47
Granovetter, Mark 13
Gras, Alain 149
Griesemer, James 78–9, 160
Griliches, Zvi 7, 10
Gross, Neal 9–10
Guterl, Fred 162

Hall, Stuart 66–7, 69
harnessing example 23–4
Haudricourt, André-Georges 15
Hawkes, Terence 59
Hawkins, Richard 148

Heilbroner, Robert L. 25
Hennion, Antoine 154
Heritage, John C. 74
Herskovits, Melville 45–6
Hert, Philippe 69
Hiltz, Starr Roxane 139
Hirsch, Eric 68
Hoddeson, Lilian 65
Hoffsaes, Colette 33
holography example 22
Homebrew Computer Club 136
Hounshell, David A. 88, 146
Hughes, Thomas 26, 84, 104, 112, 132, 157
Hugo, Victor 127
hybrid corn studies 7, 9–10

ICT studies
  business environments 70–71
  ethnomethodology 75–6
  uses 68–70
  see also information technology, uneconomic adoption
imaginaire
  discourse examples 130–33, 138–43
  genealogies of 120–23
  producers 126–30
  role in technological development 133–5
  slow development 123–4
  social *imaginaire* and frame of use 125–6, 149–52
  technological imaginaire and frame of functioning 145–9
incorporation of ICT objects 68
incremental innovation 114
influence 121–2
information technology, uneconomic adoption 134
'innerness' of technical objects 86
Innis, Harold 27–8
innovation
  analysis 163–5
  definitions 6–7, 55
  phases 157–62
innovation taxonomy 114–15
innovation users 153
Intel 136
interactionist sociology 76–9
Internet *imaginaire* 138–43

interpersonal contact 11
interplanetary communication 128–9
Inuit example 45
Invention and Economic Growth (Schmookler) 20
Isambert, François-André 72

Jennings, Tom 140
Jobs, Steve 136
Jouët, Josiane 48, 65, 68–9, 82, 93, 161

Katz, Elihu 11, 67
Kline, Ronald 69, 84
Kling, Rob 140
Knight, Kenneth 110
Kondratiev, Nikolai 116–17
Kroeber, Alfred 25
Kuhn, Thomas 48, 112

La Science et ses réseaux (Callon) 54
La Vie électrique (Robida) 127
Laboratory Life, The Construction of Scientific Facts (Latour and Woolgar) 53, 63–4
Lacaze, Dominique 127
Landes, David 5
Latour, Bruno 3, 53–61, 147
Laughlin, Charles 47
Lazarsfeld, Paul 27
Le Bas, Christian 9, 18
Le Chateau des Carpathes (Verne) 127
Le Travail (Zola) 128
Le Vingtième Siècle (De Robida) 127
learning by doing 8, 104
learning by using 8, 103
Lefebvre des Noëttes, Richard 23
Leroi-Gourhan, André 43–6
Les Misérables (Hugo) 127
letter sender example 81–2
Lévy, Pierre 40
Levy, Steven 135, 144
Licklider, Joseph 138–9
Lievrouw, Leah 119
Linhart, Danièle 32
literature *see* science fiction
Livet, Pierre 94
Livingston, Eric 74
Livingstone, Sonia 68
lock-ins *see* non-optimal lock-ins; technical lock-in

long cycles 116–19
Looking Backward (Bellamy) 126
Lorenz, Konrad 155
Lynch, Michael 72, 74, 96

Mackay, Hugh 66
Malerba, Franco 22
Mallet, Serge 32
Mangematin, Vincent 60
Mansfield, Edwin 7
Marchand, Marie 162
Marconi, Guglielmo 117, 129, 146, 151, 153, 161
market characteristics, as mediating technology and organization 38–9
Marshall, Alfred 101
Marti, Bernard 148
Martin, Michèle 103
Marvin, Carolyn 124, 130–31
Marx, Karl 3, 31, 101
Marx, Leo 130–31
material culture 46
matterism 28
Maunoury, Jean-Louis 18
Mauss, Marcel 43–4, 163
McCloskey, Donald 109
McGuire, J.E. 72
McLuhan, Marshall 27
Mead, George H. 86
Méadel, Cécile 151
media, sociology of uses 67–70
mediology 28
Melia, T. 72
Mendelbaum, Maurice 65
Mensch, Gerhard 22, 116–17
Midler, Christophe 63, 162
milieu extérieur 45
milieu intérieur 44–5
milieu technique 45
Minitel examples 69, 93, 161
Misa, Thomas J. 25–6
Monjardet, Dominique 38
Montgolfier, Joseph de 146
Moore, Fred 138
Morley, David 67
Moscovici, Serge 31
motor car example 82–5, 146–7
see also Citroën 2CV example
Mowery, David 19
Mueller, Milton 119

Mumford, Lewis 143–4

Nadar (pseudonym of Gaspard-Félix Tournachon) 127
Naville, Pierre 40
Nelson, Richard 13, 18, 110, 114
network externalities 103–4
The Network Nation, Human Communication via Computer (Hiltz and Turoff) 139
networks of influence 10
see also technoscience networks
Networks of Power (Hughes) 104
Noble, David 34–5
non-optimal lock-ins 106–7
nuclear reactors example 106–7
numerical control machine tools examples 34–5

object worlds 79
objectification 68
Odasz, Frank 140
Ogburn, William 25
operative sequence 44
opinion leaders, distinction from first adopters 12
optical pulsar discovery study 74
Orlikowski, Wanda 70–71, 81
Orsenigo, Luigi 110–11, 113
Ortsmann, Oscar 40
Oudshoorn, Nelly 72
Ozouf, Jacques 29

Pacey, Arnold 71
Painting and Experience in Fifteenth Century Italy (Baxandall) 154
Pasquier, Dominique 69
Pasteur, Louis 55
Patterns of Intention (Baxandall) 154
Peaucelle, Jean-Louis 33
The People's Choice (Lazarsfeld) 27
People's Computer Company 135
Perez, Carlota 114
Perriault, Jacques 69, 120–21
Perrin, Jacques 37–8, 118
Perrot, Michelle 38
Personal Computer *imaginaire* 135–8
Pestre, Dominique 71
Pharo, Patrick 73
Pinch, Trevor 49–52, 71–2, 80, 84

plans, definitions 75–6
Poirier, Jean 71
The Poverty of Philosophy (Marx) 31
Preece, William 129
printing press, effects 28–30
process innovation
 distinction from product innovation 6
 as having predictable demand 20
product information 11
product innovation, distinction from process innovation 6
Proulx, Serge 94

Quéré, Louis 51, 64, 73, 86
Quirk, John 132
qwerty keyboard example 106

radical innovation 114
railway carriage example 155
Rallet, Alain 94
Ralph 124C 41+ (Gernsback) 129
re-invention 13
Real, Bernard 6
reference groups 77
rejecters 69
repairs 84
resisters 69
Resource One 135
retribalization 27
Reynolds, Barrie 46–7
Rheingold, Howard 139, 141, 143
Ricardo, David 3
Rice, Ronald 13
Ricoeur, Paul 75, 99–100
Rip, Arie 134, 162
Robbins, Lionel 4
Robida, Albert 127
Robinson, Mike 76, 85–6
Rogers, Everett 10–11, 13, 18, 119
Roosevelt, Theodore 132
Rosenberg, Nathan 8, 14, 19, 21, 103–4, 110
Rosier, Bernard 107, 119, 162
Roszack, Theodore 137
Ryan, Bryce 10

Sadoul, Jacques 144
Sahal, Devendra 114
Sappho project 23, 90

Schaffer, Simon 49
Schmookler, Jacob 20
Schumpeter, Joseph 5–6, 102
Schütz, Alfred 73
science, as organizing work 37–8
science fiction 126–30
Science in Action (Latour) 54
science in-the-making
  as negotiation 48–9
  as rhetorical 53
SCOT (Social Construction of Technology) 49–50
script 66–7
Segal, Howard 126
Segrestin, Denis 156
Serres, Michel 72
Shapin, Steve 49
Shapiro, Carl 103, 106
Shibutani, Tamotsu 77
Silverstone, Roger 67–8
Simondon, Gilbert 92, 111–12, 147
simultaneity of inventions 25
Singer, Charles 18
Smith, Adam 3
social constructivism approach *see* technology, as social artifact
social worlds 77–8
socio-technical frames 79–87, 152–7, 161–2
socio-technical history, definitions 100–101
socio-technical systems school 32
Sol 136
solar house example 79
Solow, Robert 5
spokesperson metaphor 54–5
SST (Social Shaping of Technology) 49
Star, Susan Leigh 76, 78–9, 160
steam engine *imaginaire* 131
stirrup example 23
strategic actors 60
strategists 87–91
Strauss, Anselm 77
strong programmes 48
Suchman, Lucy 75, 96
supply–push theories 21–3
Sur la pierre blanche (France) 128

tacticians 87–8, 92–6

Tarde, Gabriel 128
Taylor, Robert 32, 37–8
Taylorism 32, 37–8
technical convergence 45
technological change 5
technological competition 101–9
technological complementarities 104–5
technological controversies 49–50
technological determinism, and the organization of labour 31–40
technological frameworks 49
technological lock-in 105–6
  see also concretization
technological moments 26
technological paradigms 112–15
technological revolutions 114–15
technology
  as cultural transformation 15–18
  historians' analysis 13–15
  as a language 47
  as missing factor of production 3–7
  as social artifact 48–53
technology analysis model
  frames of reference 80–87, 123–4
  objectives 79–80
  socio-technical action
    during innovation 87–91
    during stability 91–6
technoscience networks
  actors 58–60
  context 60–62
  limitations 62–5
  networks 57–8
  overview 53–7
telegraphy examples 28, 84, 90, 128, 134, 150–52
telephony examples 83–4, 150, 151
Tesla, Nicolas 129
Thévenot, Laurent 94, 99
Thomas, Dorothy 25
threshold models 13
Thuillier, Pierre 72
time passage
  as evolutionary 109–12
  long cycles 116–19
  as multiple paradigms 112–19
  technological competition 101–9
  viewed as history 99–101
Toda example 45–6
tools, and gestures 44

Touraine, Alain 56
Tournachon, Gaspard-Félix *see* Nadar
triode example 64–5
Turkle, Sherry 68
Turoff, Murray 139
Twenty Thousand Leagues under the Sea (Verne) 127
'two-step flow' thesis 27

user-strategists 90–91
users' roles
　adjusting technical objects 92–4
　cooperation 94–6
　during innovation 87–91
　in information technologies 70–71
　overview 65–6
　semiotic approaches 66–7
uses, sociology of 65, 67–70
Usher, Abbott P. 25
Utterback, James 19, 161

Valente, Thomas W. 13
Van Lente, Harro 134, 162
Varian, Hal 103, 106
VCR example 93
Velkovska, Julia 69, 93
Vercruysse, Jean-Pierre 150
Verhoest, Pascal 150
Verne, Jules 126–7
Veyne, Paul 100
videophone examples 95

Villiers de L'Isle-Adam, Auguste, de 127
The Virtual Community (Rheingold) 141
Von Hayek, Friedrich 4
Von Hippel, Eric 90, 153, 156
von Siemens, Werner 131

Walras, Léon 6
water mill example 24
weak determinism 25
Weber, Max 100
Webster, Juliet 35
The Well 139–41
Wheelwright, Steven 79
White, Lynn 23
Williams, Raymond 123
Williams, Robin 35, 49
Williams, Rosalind 130, 132–3, 144
Winner, Langdon 52
Winter, Sidney 13, 110, 114
Wired (journal) 143
Woolgar, Steve 53, 60, 64, 66, 72
Wozniak, Steve 136
Wyatt, Sally 69

Yates, JoAnne 70, 91, 134

Zitt, Michel 20
Zola, Émile 128
zoology museum study 78–9